Gum Karaya

Polysaccharide gums have a wide range of applications due to their hydrophilic properties. They have been used as a gelling agent, encapsulating agent, thickener, emulsifier, and stabilizer. Sourcing natural gums from botanical and plant sources has become an important focus in producing acceptable food ingredients in liquids and semi-solid forms. This is mainly because of the positive attitude of consumers toward plant-based gums rather than other gums from animal and microbial sources.

Gum karaya, also known as Indian tragacanth, is a vegetable gum produced as an exudate by trees of the genus *Sterculia.* Gum karaya has very strong swelling properties, high viscosity, and very poor solubility because of its acetyl groups. Therefore, it is mainly used in cosmetics and pharmaceuticals. Gum karaya is one of the least-soluble gums used for many industries, such as petroleum and gas, textile, paper and pulp, leather and allied products, ammunition and explosives, electrical appliances, adhesives, confectionery, medicine, pharmaceuticals, and cosmetics.

Because of the crude tapping method and overexploitation, the population of karaya trees has markedly declined. In the absence of cultivation of this tree on the regular plantation, there is grave concern about the loss of wild germplasm of *S. aurens.* As gum karaya is vital for the tribal economy and its trade value is substantial, there is a pressing need to develop a scientific and sustainable tapping method to increase the yield and ensure the survival of the tapped trees. There is also a need for a large-scale plantation of gum karaya. People are becoming health-conscious and are taking natural products as food ingredients. This book will be useful to all the people interested in natural food additives.

- This book gives the details of gum karaya–yielding species and their distribution.
- Scientific methods of gum tapping are given for sustainable gum production.
- Propagation methods, both field and in vitro, are given.
- Uses of gum karaya, in cosmetics, pharmaceuticals, and food, are given.
- Genetic diversity and biotechnological intervention for the improvement of gum karaya–yielding species are included.
- Threats and conservation of gum karaya–yielding species are given.

Gum Karaya

Botany, Gum Tapping, Applications, and Biotechnology

Edited by T. Pullaiah, Niranjan Prasad,
Lepakshi Md. Bhakshu,
and Mallappa Kumara Swamy

CRC Press
Taylor & Francis Group
Boca Raton London New York

CRC Press is an imprint of the
Taylor & Francis Group, an **informa** business

Cover design: Kothareddy Prasad

First edition published 2025
by CRC Press
2385 NW Executive Center Drive, Suite 320, Boca Raton FL 33431

and by CRC Press
4 Park Square, Milton Park, Abingdon, Oxon, OX14 4RN

CRC Press is an imprint of Taylor & Francis Group, LLC

ISBN: 978-1-032-57343-4 (hbk)
ISBN: 978-1-032-57346-5 (pbk)
ISBN: 978-1-003-43896-0 (ebk)

DOI: 10.1201/9781003438960

Typeset in Times
by Apex CoVantage, LLC

Contents

Preface

GUM KARAYA– A TREASURE FROM THE *STERCULIA*: FROM ANCIENT HEALING TO MODERN APPLICATIONS

Gum karaya is a fascinating natural exudate brimming with history, therapeutic potential, and diverse industrial applications. This book unveils the secrets of this versatile gum, its intimate connection to the mighty *Sterculia* trees, and its journey from ancient Ayurvedic remedies to a cornerstone of modern science.

The present book delves into the ancient lore, tracing gum karaya's usage in traditional medicine across cultures. Imagine Indian sages harnessing its soothing properties for coughs and inflammation, while Arabian pharmacopoeias praised its laxative effects. Witness its evolution into a mainstay of Ayurvedic Materia medica, earning the moniker 'Indian tragacanth'. Pharmacognosy unveils the hidden treasures within gum karaya. We dissect its intricate chemical composition, delving into the world of polysaccharides, galacturonic acid, and rhamnose. Witness the scientific validation of its age-old uses, as clinical trials reveal its efficacy in treating ailments like constipation and gastrointestinal disorders.

The book has focused on the realm of phytochemistry to disclose the chemical properties and its diversity of secondary metabolites within the *Sterculia* plants, the factors influencing gum production, and the sustainable tapping methods that safeguard these precious trees. It is helpful in learning how the modern tissue culture techniques offer hope for enhanced gum yield as well as its conservation efforts.

Unveiling the versatility of gum karaya beyond the realm of medicine, this book witnesses its transformative ability in industries such as textiles, food, oil drilling, and cosmetics. It is explored with a role as a thickener, emulsifier, and adhesive, enhancing texture and stability in everyday products. The gum karaya supports the life of poor people around its vicinity; as a raw material, it supports the industrial corridor and plays a vital role in the country's economy, proving its global value, as explained in various chapters.

As we explore the benefits of gum karaya, we cannot ignore its vulnerability and the solution for its improvement. The challenges facing *Sterculia* trees, threats like overexploitation and deforestation, are leading to its disappearance. In this connection, the book reviews the sustainable practices and conservation strategies that ensure this natural treasure thrives for generations to come. This book serves as a guide to a sustainable future, where tradition meets innovation and a humble 'exudate' continues to enrich our lives in countless ways.

T. Pullaiah
Niranjan Prasad
Lepakshi Md. Bhakshu
Mallappa Kumara Swamy

About the Editors

Prof. T. Pullaiah obtained his MSc and PhD degrees in Botany from Andhra University. He was a post-doctoral fellow at Moscow State University, Russia, during 1976–1978. He travelled widely in Europe and America and visited universities and botanical gardens in about 17 countries. He joined Sri Krishnadevaraya University as a lecturer in 1979 and became a professor in 1993. He has held several positions at the university, including as dean, faculty of biosciences, head of the Department of Botany, head of the Department of Biotechnology, and member of the Academic Senate. He was the president of the Indian Botanical Society (2014), the president of the Indian Association for Angiosperm Taxonomy (2013), and a fellow of the Andhra Pradesh Academy of Sciences. Under his guidance, 54 students obtained their doctoral degrees. He has authored 70 books, edited 40 books, and published over 340 research papers. His books have been published by reputed international publishers like CRC Press, Springer, Elsevier, Wiley Blackwell, Apple Academic Press, etc. He was also a member of the Species Survival Commission of the International Union for Conservation of Nature (IUCN).

Dr Niranjan Prasad did his BE (agricultural engineering) from CTAE, Udaipur (1988), and MTech (1991) and PhD from IIT, Kharagpur (1996). He obtained his post-doctoral training in smart/active packaging from Michigan State University, USA. He worked as a lecturer at NERIST, Nirjuli, Arunachal Pradesh, during 1996–1997. Later, he joined Indian Lac Research Institute, Ranchi, as an agricultural research service scientist in 1997 and presently works as a principal scientist and the head of Agri Bio-resources Augmentation Division at ICAR-National Institute of Secondary Agriculture, Ranchi, India. He is also working as a project coordinator of the All India Network Project on Harvesting Processing and Value Addition of Natural Resins and Gums. He worked as a consortium principal investigator of the National Agricultural Innovation Project (World Bank) on Lac Value Chain. During his 28 years of experience in research and extension, he published 65 research papers in national and international journals. He has developed several tools and machines for lac cultivation and processing which have been commercialized. Dr Prasad is a member of several professional societies. He is the recipient of many awards, namely, the ISAE Jain Irrigation Award (1987–1988); DST SERC Visiting Fellowship (1999); ISAE Distinguished Service Certificate (2002–2003); IINRG Ranchi Certificate of Merit (2004–2009); ISAE Commendation Medal 2008–2009; Bioved, Allahabad Fellowship 2012; Fellow of ISAE (2019), and David Foundation Award (2020). Dr Prasad was also nominated as a member of the Institute Management Committee of ICAR-CIPHET, Ludhiana; ICAR-CIAE, Bhopal; ICAR-NIRJAFT, Kolkata; and ICAR-IASRI, New Delhi.

Dr Lepakshi Md. Bhakshu completed his master's degree in botany (1998) from Sri Venkateswara University, Tirupati, and PhD in Botany (2002) from Sri Krishnadevaraya University, Anathapuramu. He has 15 years of teaching and 20 years of research experience in ethnobotany, ethnopharmacology, phytochemical studies, and biodiversity conservation. He published 24 research publications in national and international journals, presented 8 research papers in international and national conferences, and participated in 20 national and international symposia besides writing 20 book chapters. He received the two best posters in the national-level symposium. He worked as a project fellow at EPTRI, Hyderabad, and a co-author of the *Proceedings of the National Seminar on the Conservation of Eastern Ghats*, published by the EPTRI, Hyderabad. He is the recipient of a post-doctoral fellowship and is a Dr DS Kothari national fellow (UGC) at the University of Hyderabad, Telangana. Presently, he is working as an assistant professor at the Department of Botany, Dr YSR Government Degree College, Vedurukuppam, Chittoor (Dt), Andhra Pradesh, India.

Dr Mallappa Kumara Swamy is a professor and the head of the Department of Biotechnology at East West First Grade College, Bengaluru, India. He completed his post-doctoral research at the Department of Crop Science, Faculty of Agriculture, University Putra Malaysia (UPM), Serdang, Malaysia. Before that, he had worked as an associate professor and the head of the Department of Biotechnology, Padmashree Institute of Management and Sciences, Bangalore University, Bengaluru, India. He received his PhD (biotechnology) from Acharya Nagarjuna University, Guntur, India, in 2013. He has more than 17 years of teaching and research experience in the fields of plant biotechnology, secondary metabolite production, phytochemistry, and bioactive studies. Dr Swamy has authored 65 research publications in peer-reviewed journals and 68 book chapters. Recently, he has edited 12 books published by Springer Nature Singapore Pvt. Ltd., Singapore; 1 book by CRC Press LCC, USA; 1 book by Studium Press Pvt. Ltd., India; and 1 edited book by Elsevier, USA. He also served as an editorial board member and reviewer for several high-impact factor international journals. His current research is focused on cell and tissue culture technology, phytochemistry, phytocompounds isolation, production and their biological evaluation, and nanotechnology for medical applications.

Contributors

Ajay Kumar
Department of Plant Science
School of Biological Sciences
Central University of Kerala
Kasaragodu 671316
India

Boregowda Purushotham
Department of Biotechnology
East West First Grade College of Science
Bengaluru 560091
Karnataka
India

Abhijit Dey
Department of Biotechnology
School of Engineering and Technology (SET) Sharda University
Greater Noida
India

Lepakshi Md. Bhakshu
Door #4–1595, Kolimi Street
Greamspet Chittoor, 517002
Andhra Pradesh
India

Mallappa Kumara Swamy
NO. 84, 2nd Cross
Nisarga Layout
Near Ullal RTO Office
Bengaluru 560091
India

Niranjan Prasad
ICAR-National Institute of Secondary Agriculture
Ranchi – 834 010
India

S. Noorunnisa Begum
Associate Professor
74/2, Jarakabande Kaval
Post Attur
via Yelahanka
Bengaluru 560064
Karnataka
India

T. Pullaiah
6–3–244, Ramnagar Extension
Anantapur 515004
Andhra Pradesh
India

R. Patturaj
Scholar, 74/2, Jarakabande Kaval
Post Attur
via Yelahanka
Bengaluru 560064
Karnataka
India

K. Ravikumar
Emeritus Professor
74/2, Jarakabande Kaval
Post Attur
via Yelahanka
Bengaluru 560064
Karnataka
India

K. Venkata Ratnam
Department of Botany
Rayalaseema University
Kurnool
Andhra Pradesh
India

1 Gum Karaya – The Green Gold

T. Pullaiah

1.1 INTRODUCTION

Polysaccharide gums have a wide range of applications due to their hydrophilic properties. They have many applications as a gelling agent, encapsulating agent, thickener, emulsifier, and stabilizer (Williams and Phillips, 2000; Patel and Goyal, 2015). In the food industry, they are mainly used to modify the physicochemical and functional properties (i.e. rheological properties, solubility, oil- and water-holding capacity) of food products (Nussinovitch, 1997; Amid and Mirhosseini, 2012). Sourcing natural gums from botanical and plant sources has become an important focus to produce acceptable ingredients in liquids and semi-solid foods (Rana et al., 2011). This is mainly because of the positive attitude of consumers toward plant-based gums rather than other gums from animal and microbial sources (Nishinari et al., 2000).

Gum karaya, also known as Indian tragacanth, is a vegetable gum produced as an exudate by trees of the genus *Sterculia*. Gum karaya (GK) is a commercially viable polysaccharide gum extracted from the *Sterculia urens* tree, which is extensively grown in India (Verbeken et al., 2003). It is an exudate comprised of 60% ordinary sugar dregs (galactose and rhamnose), 40% acidic dregs (galacturonic and glucuronic acid), and 8% acetyl groups (Morris, 1995). Chemically, gum karaya is an acid polysaccharide composed of sugars galactose, rhamnose, and glucuronic and galacturonic acid. Gum karaya is obtained mostly from Indian plantations of *Sterculia urens* and smaller plantations of *Sterculia villosa*. In Sudan and elsewhere in Africa, Gum karaya can be obtained from *Strculia setosa*. Karaya gum has been reported to be collected from *S. urceolata* and *S. foetida* in Indonesia, *S. setigera* in Africa, and from *S. caudata* in Australia (Gautami and Bhat, 1992). Commercial tapping of gum karaya is done by blazing, by peeling, or by making deep cut in the base of the tree trunk with an axe or sickle. Trees of *Sterculia quinqueloba* are tapped for the gum, which is traded as 'gum karaya' together with the gum of other *Sterculia* spp. (*www.prota4u.org/database/protav8.asp?g=psk&p=Sterculia%20quinqueloba*).

Gum karaya has very strong swelling properties, high viscosity, and very poor solubility because of its acetyl groups; therefore, it is mainly used in cosmetics and pharmaceuticals rather than in food processes (Morris, 1995; Bharadwaj et al., 2000; Weiping et al., 2000; Silva et al., 2003). Gum karaya is one of the least-soluble gums used for many industries, such as petroleum and gas, textile, paper and pulp, leather and allied products, ammunition and explosives, electrical appliances, adhesives, confectionery, medicine, pharmaceuticals, and cosmetics (Gautami and Bhat, 1992).

DOI: 10.1201/9781003438960-1

FIGURE 1.1 *Sterculia urens* tree.

Source: Image by Dr N. Venugopal with permission.

World production of gum karaya is currently estimated at about 5,500 t per annum and is on a declining trend. India is the only regular producer overwhelmingly dominating international trade in this gum. Nowadays, the gum karaya price fluctuates in the range US$2,250–6,000/t according to grade. Fair average quality (FAQ) gum is about US $3,000/t. Yield of this gum has shown a phenomenal decrease from 6,838 t in 1975–1976 to 461.3 t in 1990–1991. During this period, the price has shot up Rs. 7,491 per ton to Rs. 83,361 (Gautami and Bhat, 1992). For 100 years, India has been the largest producer and exporter of karaya gum. India produces 20,000 t of exudate gums, in which gum karaya alone contributes about 15,000 t (Yadav and Basera, 2013).

Carefully harvested and graded gum of *Sterculia quinqueloba* meets the requirements of commercial 'gum karaya' (normally derived from the Indian species *Sterculia urens* Roxb. and having the European code E416): a maximum of 200 g/kg moisture content and 30 g/kg of insoluble matter; it also has low levels of aluminium, manganese, and cobalt. The high tannin content of the gum of *Sterculia quinqueloba*, however, limits its usefulness in foods and pharmaceuticals. The gum of *Sterculia quinqueloba* possesses solubility and viscosity properties similar to commercial gum arabic from *Acacia* spp., especially in having a rather low viscosity which is similar in hot and cold water. However, *Sterculia quinqueloba* gum does not meet all specifications required of gum arabic, and much higher amounts of the gum are needed to obtain similar results (*www.prota4u.org/database/protav8.asp?g=psk&p=Sterculia %20quinqueloba*).

Increasing human population enjoyed it, and the growing demand fell into search for other source of gum karaya. The global demand and market lobby started threatening the gum karaya existence. This is felt as a threat to the livelihood of the relying communities and export revenue of the developing countries. Increasing human population enjoyed it, and the growing demand fell into a search for other sources of gum karaya.

FIGURE 1.2 Dr N. Venugopal (standing) giving training to the tribals in scientific extraction of gum karaya in Central Madhya Pradesh, India.

Source: Image courtesy of Dr N. Venugopal, with permission.

On account of crude tapping method and overexploitation, the population of karaya trees has markedly declined. In absence of cultivation of this tree in regular plantation, there is grave concern about the loss of wild germplasm of *S. urens* (Figures 1.1 & 1.2). International trade of gum karaya is exclusively indigenous to India and is vital for the tribal economy. As gum karaya is vital for the tribal economy and its trade value is substantial, there is a pressing need to develop a scientific and sustainable tapping method to increase the yield and to ensure the survival of the tapped trees. There is also need for large-scale plantation of gum karaya. There is a need for assessment of genetic diversity and biotechnological intervention for improvement of gum karaya–yielding species.

Hence, there is an urgent need for the assessment of the species yielding gum karaya. There is also a necessity for the scientific methods of extraction and propagation of the trees. A review of the applications is also necessary. Biotechnological applications of the species of *Sterculia* especially *Sterculia urens* are needed for improvement of yield of Gum karaya. This book is an answer for all these.

REFERENCES

Amid, B.T. and H. Mirhosseini. 2012. Optimization of aqueous extraction of gum from Durian (*Durio zibethinus*) seed: A potential, low cost source of hydrocolloid. *Food Chem.* 132: 1258–1268

Bhardwaj, T.R., M. Kanwar, R. Lal and A. Gupta. 2000. Natural gums and modified natural gums as sustained-release carriers. *Drug Develop. Indus. Pharm.* 26: 1025–1038

Gautami, S. and R.V. Bhat. 1992. *A monograph of gum Karaya.* National Institute of Nutrition, ICMR, Hyderabad

Morris, E.R. 1995. *Food polysaccharides and their applications*; Stephens, A.M.; Ed.; Marcel Dekker, New York, NY, pp. 341–375

Nishinari, K., H. Zhang and S. Ikeda. 2000. Hydrocolloid gels of polysaccharides and proteins. *Curr. Opin. Coll. Interf.* 5: 195–201

Nussinovitch, A. 1997. *Hydrocolloid applications.* Blackie Academic and Professional, London, pp. 140–151.

Patel, S. and A. Goyal. 2015. Applications of natural polymer gum Arabic: A review. *Intern. J. Food Properties* 18: 986–998

Rana, V., P. Rai, A.K. Tiwary, R.S. Singh, J.F. Kennedy and C.J. Knill. 2011. Modified gums: Approaches and applications in drug delivery. *Carbohyd. Polym.* 83: 1031–1047

Silva, D., A. Brito, R. De Paula, J. Feitosa and H. Paula. 2003. Effect of mono and divalent salts on gelation of native, NA and deacetylated *Sterculia striata* and *Sterculia urens* polysaccharide gels. *Carbohyd. Polym.* 54: 229–236.

Verbeken, D., S. Dierckx and K. Dewettinck. 2003. Exudate gums: Occurrence, production, and applications. *Appl. Microbiol. Biotechnol.* 63: 10–21

Weiping, W., G. Phillips and P. Williams. 2000. Tragacanth; Karaya. In: Philips, G.O. and Williams, P. (eds.) *Handbook of hydrocolloids.* CRC Press, New York, pp. 231–245.

Williams, P.A. and G.P. Phillips. 2000. *Handbook of hydrocolloids.* CRC Press, New York, pp. 1–14

Yadav, M. and K. Basera. 2013. *Status of forest products production and trade* (Indian Institute of Forest Management Working Paper Series (2013/1)). Indian Institute of Forest Management, Hyderabad, p. 14

2 Botany of Gum Karaya–Yielding Species

T. Pullaiah

2.1 INTRODUCTION

Sterculia is a genus of flowering plants in the family Malvaceae, subfamily Sterculioideae (previously placed in the now-obsolete Sterculiaceae). Members of the genus are colloquially known as tropical chestnuts. The scientific name of *Sterculia* is taken from *Sterculius* of Roman mythology, who was the god of manure; this is in reference to the unpleasant smell of the flowers of this genus (e.g., *Sterculia foetida*).

The Plant List includes 439 scientific plant names of species rank for the genus *Sterculia*. Of these, 60 are accepted species names (www.theplantlist.org/browse/A/Malvaceae/Sterculia/). These are:

Sterculia abbreviata E.L.Taylor ex Mondragon
Sterculia africana (Lour.) Fiori
Sterculia allenii E.L.Taylor
Sterculia amazonica E.L.Taylor ex Mondragon
Sterculia apeibophylla Ducke
Sterculia apetala (Jacq.) H. Karst
Sterculia brevissima H.H.Hsue
Sterculia caribaea R.Br.
Sterculia ceramica R.Br.
Sterculia cinnamomifolia Tsai & Mao
Sterculia colombiana Sprague
Sterculia corrugata Little
Sterculia costaricana Pittier
Sterculia dawei Sprague
Sterculia euosma W.W.Sm.
Sterculia excelsa Mart
Sterculia foetida L.
Sterculia frondosa Rich.
Sterculia gengmaensis H.H.Hsue
Sterculia guangxiensis S.J.Xu & P.T.Li
Sterculia guapayensis Cuatrec.
Sterculia hainanesis Merr. & Chun
Sterculia henryi Hemsl.
Sterculia hymenocalyx K.Schum.

DOI: 10.1201/9781003438960-2

Sterculia impressinervis H.H.Hsue
Sterculia kavae P.E.Berry
Sterculia kingtungensis H.H.Hsue
Sterculia lanceifolia Roxb.
Sterculia lanceolata Cav.
Sterculia mexicana R.Br.
Sterculia micrantha Chun & H.H.Hsue
Sterculia mirabilis (A.Chev.) Roberty
Sterculia monosperma Vent.
Sterculia peruviana (D.R.Simpson) E.L.Taylor ex Brako & Zarucchi
Sterculia pexa Pierre
Sterculia pinbienensis Tsai & Mao
Sterculia principis Gagnep.
Strculia pruriens (Aubl.) K.Schum.
Sterculia purpurea Exell
Sterculia quinqueloba (Garcke) K.Schum.
Sterculia rugosa R.Br.
Sterculia scandens Hemsl.
Sterculia setigera Delile
Sterculia simaoensis Y.Y.Qian
Sterculia speciosa K.Schum.
Sterculia steyermarkii E.L.Taylor ex Mondragon
Sterculia stipulifera Ducke
Sterculia striata A. St.-Hil. & Naudin
Sterculia subnobilis H.H.Hsue
Strculia subracemosa Chun & H.H.Hsue
Sterculia subviolacea K.Schum.
Sterculia tavia Baill.
Sterculia tessmannii Mildbr.
Sterculia tonkinensis Aug. DC.
Sterculia tragacantha Lindl.
Sterculia venezuelensis Pittier
Sterculia villifera Steudl
Sterculia villosa Roxb.
Sterculia xolocotzii T. Wendt. & E.L.Taylor
Sterculia yuanjiangensis H.H.Hsue & S.J.Xu

In tropical Africa, about 25 species of *Sterculia* can be found.

STERCULIA L.

Sterculia balanghas L., Sp. Pl. 1007. 1753. var. ***glabrescens*** Masters in Hook. f., Fl. Brit. India 1: 358. 1874; Malick in B.D. Sharma & Sanjappa, Fl. India 3: 458. 1993.

Trees; young parts rusty tomentose. Leaves simple, crowded at the ends of branches, oblong-ovate; petioles 3–5 cm long. Flowers yellow or greenish-purple, in axillary or terminal, rusty-tomentose panicles, scented; calyx campanulate, 5-lobed; stellate-hairy outside, densely along the margins, hispid inside. Male flowers: staminal column c. 2 mm long. Female flowers: stigma curved, as long as styles. Follicles

4–5, horizontally spreading, woody, covered with rusty tomentum; seeds oblong or ovoid, black, shining.

Throughout hotter parts of India, Sri Lanka, and Myanmar.

Fl.: September to December; Fr.: April to June.

Sterculia cordata Blume Bijdr. 83. 1825; Malick in B.D. Sharma & Sanjappa, Fl. India 3: 459. 1993. *S. pubescens* Masters in Hook. f., Fl. Brit. India 1: 357. 1874.

Deciduous trees, up to 10 m high, bark and petioles rusty velutinous. Leaves simple elliptic-ovate or obovate, glabrous above and rusty velutinous beneath; petioles 2–18 cm long. Flowers unisexual in axillary panicle. Male flowers: Calyx campanulate, hairy outside; stamens ca 10 on a column. Female flowers: calyx broadly campanulate, sparsely outside, densely inside; ovary obscurely 5-lobed, stellate-hairy.

India in Andaman and Nicobar Islands.

Fl.: October.

Sterculia foetida L., Sp. Pl. 1008. 1753; Masters in Hook. f., Fl. Brit. India 1: 354. 1874; Malick in B.D. Sharma & Sanjappa, Fl. India 3: 459. 1993.

Vern.: Bengali: Badam, Jangli-Badam; Kannada: Patala mara, Penari; Marathi: Jungli-Badam, Pun; Tamil: Pinari, Pottaikavalam; Kudiraippiduku.

Large deciduous trees, up to 30 m high; bark whitish, flaking off; branches whorled, horizontal. Leaves digitately lobed, petioles 14–24 cm long, leaflets 5 or 9, elliptic, 8–13 × 3–5 cm, glabrous, base cuneate, margin entire, apex acute or caudate, pinnately veined; petiolules up to 1 cm long; stipules ensiform, caducous. Flowers dull orange-coloured, in axillary panicles; calyx campanulate, deeply 5-parted, lobes oblong-lanceolate, spreading, villous within, much longer than the tube. Male flowers: anthers 10–15 on ca 1.2 cm long staminal column. Female flowers: carpels 5, ovaries on 5–7 mm long androgynophore, hairy, surrounded by sterile anthers at base, stigmas as many as carpels, radiating. Follicles 1–5, boat-shaped, glabrous, woody, beaked; seeds 10–15, ovoid-ellipsoid, black, smooth.

Planted in gardens, along roadsides, house yards, office fences as an avenue and ornamental tree.

Fl.: March to April; Fr.: May to August.

Sterculia guttata Roxb. Fl. Ind. 3: 148. 1832; Masters in Hook. f., Fl. Brit. India 1: 355. 1874; Malick in B.D. Sharma & Sanjappa, Fl. India 3: 462. 1993.

Vern.: Assamese: Hirikh; Kannada: Jenu Kathala, Hulitarady mar; Malayalam: Kithondi, Kavalam; Tamil: Kavali, Thonidi.

Deciduous trees, up to 20 m high, trunk straight, bark much cracked, ash-coloured. Leaves simple, ovate-oblong or obovate, glabrescent above, rusty stellate-pubescent beneath; petiole 3–5 cm long. Flowers in racemose panicles, branches rusty tomentose; calyx campanulate, densely pubescent outside, glandular inside. Male flowers: anthers 10–15, on curved staminal column. Female flowers: ovaries on 4–5 mm long gynandrophore, style curved. Follicles 2–5, obovoid, coriaceous, compressed, obscurely ribbed, brown tomentose, pink inside; seeds in each follicle 3–4.

In evergreen forests in India, Sri Lanka, and Malesia.

Fl.: September to May; Fr.: February to August.

Sterculia hamiltonii (O. Kuntze) Adelb., Blumea 5: 506. 1945; Malick in B.D. Sharma & Sanjappa, Fl. India 3: 464. 1993. *Clompanus hamiltonii* O. Kuntze, Rev.

Gen. Pl. 1: 77. 1871. *Sterculia coccinea* Roxb., Fl. Ind. 3: 151. 1832 non Jack 1822; Masters in Hook. f., Fl. Brit. India 1: 357. 1874.

Vern.: Assamese: Nak-chepeta, Saglepapio, Komkelu.

Shrub or small tree, bark grey, warty. Leaves simple, elliptic-lanceolate, oblanceolate, of narrowly oblong, glabrous above, stellate hairy beneath; petiole 7–23 cm long. Flowers pale, in axillary panicles; calyx campanulate. Male flowers: staminal column curved. Female flowers: ovaries on gynandrophore, styles curved, stigmas 5-lobed. Follicles 2–5, oblong-lanceolate, beaked, velvety outside; seeds 4–8, ovoid.

India, Bhutan, and Myanmar.

Sterculia hyposticta Miq. Fl. Ind. Bat. Suppl. 399. 1861; Malick in B.D. Sharma & Sanjappa, Fl. India 3: 464. 1993.

Shrub, 1–2 m high. Leaves simple, elliptic-lanceolate or oblong; petiole 1–1.5 cm long. Flowers in few-flowered panicles; calyx globose tube. Male flowers: staminal column 1 mm long. Female flowers: ovaries globose, velvety, stigmas 5-fid, recurved. Follicles ellipsoid, covered with yellowish brown hairs outside.

India and Malesia.

Fl.: November; Fr.: February.

Sterculia khasiana King ex Debbarman in Kanjilal et al., Fl. Assam 1: 154. 1934; Malick in B.D. Sharma & Sanjappa, Fl. India 3: 466. 1993.

Tree. Leaves simple, elliptic or obovate, petiole 8–13 mm long. Flowers in terminal lax racemes; calyx woolly outside. Male flowers: staminal column thick, reflexed. Female flowers: ovaries hairy, style curved.

India, endemic to Meghalaya.

Sterculia kingii Prain, J. Asiat. Soc. Beng. 73, n.s. 295: 192. 1904; Malick in B.D. Sharma & Sanjappa, Fl. India 3: 466. 1993.

Vern.: Nepali: *Chiwaripat.*

Small tree, bark greyish. Leaves simple, broadly elliptic-ovate or lanceolate; petiole 3–4 long. Flowers yellow, in axillary panicles or racemes at the ends of drooping branchlets; calyx membranous, pilose. Male flowers: staminal column curved downwards. Female flowers: ovary on gynandrophore, style curved downwards, stigma 5-fid. Follicles 4–5, hairy outside.

India and Bhutan.

Fl. and Fr.: May to June.

Sterculia macrophylla Vent., Hort. Malm. 2, t. 91. 1805; Masters in Hook. f., Fl. Brit. India 1: 356. 1874; Malick in B.D. Sharma & Sanjappa, Fl. India 3: 468. 1993.

Trees. Leaves simple, broadly ovate to suborbicular; petiole 3–15 cm long. Flowers in deflexed panicles, rusty tomentose; calyx cup-shaped, densely stellate-hairy. Male flowers: staminal column 1–1.2 mm long. Female flowers: ovaries on short gynophore, 5-angled. Follicles 1–5, villous; seeds 1–2.

India and Myanmar.

Fl. and Fr.: May to July.

Sterculia parviflora Roxb. Fl. Ind. 3: 147. 1832; Masters in Hook. f., Fl. Brit. India 1: 359. 1874; Malick in B.D. Sharma & Sanjappa, Fl. India 3: 468. 1993.

Trees, bark light-coloured. Leaves simple, oblong, oblong-ovate, or elliptic, petiole 2–3 cm long. Flowers yellowish-brown in panicles, stellate-hairy; calyx globular, tube urceolate. Male flowers: staminal column with 10 fertile stamens. Female flowers: ovary hairy, stigma 5-fid. Follicles 3–5 cm long, oblong, beaked; seeds black.

India, Myanmar, and Malesia

Fl.: January to February.

Sterculia quinqueloba (Garcke) K.Schum., Bot. Jahrb. Syst. 15: 135. 1892. *Cola quinqueloba* Garcke, Naturw. Reise Mossambique 130. 1861.

Deciduous small to medium-sized tree, up to 20(–40) m tall; bole branchless for up to 4.5 m or more, up to 100 cm in diameter, usually straight and cylindrical; bark surface grey to pale greenish-brown, smooth, flaking, inner bark mottled red and fibrous; young twigs with triangular bud-scales. Leaves alternate, simple; stipules early caducous; petiole 6–18 cm long; blade orbicular in outline, 8–36 cm × 10–34 cm, with 5 triangular, acuminate lobes, deeply cordate at base, soft hairy with stellate hairs on lower surface. Inflorescence an axillary, narrow panicle 15–35 cm × 6–11 cm, sticky hairy; bracts 3–7 mm long, caducous. Flowers unisexual, regular, 5-merous, pale or yellowish green, c. 3.5 mm long; perianth campanulate with lobes, c. 1 mm × 1 mm, outside sticky hairy; male flowers with 8–10 anthers borne on a common stalk c. 1.5 mm long; female flowers with ovary consisting of 5 carpels united loosely. Fruit usually consisting of 5 woody follicles 5–8 cm × 1–1.5 cm, densely soft hairy, yellow-brown, sticky, many-seeded. Seeds ellipsoid, 7 mm × 4–7 mm, grey-black, glabrous (*www.prota4u.org/database/protav8.asp?g=psk&p=Sterculia%20 quinqueloba*).

Sterculia roxburghii Wallich, Pl. Asiat. Rar. 3: t. 262. 1832; Masters in Hook. f., Fl. Brit. India 1: 356. 1874; Malick in B.D. Sharma & Sanjappa, Fl. India 3: 468. 1993.

Vern.: Assamese: Nag-phona, Nag phena; Garo: Mimong-omak, Misi-chik-udari.

Tree, bark dark-brown to ash grey. Leaves simple, ovate, obovate, or oblanceolate, petiole 2–7 cm long. Flowers in axillary panicles; calyx scarlet, campanulate. Male flowers: staminal column 1 mm long. Female flowers: ovary 1 mm long, stigmas 5. Follicles 2–5; seeds 4–8 in each follicle, black, shining.

India and Bhutan.

Fl.: February to September; Fr.: March to June.

Sterculia rubiginosa Vent., Hort. Malm. 2. t. 91. 1805; Masters in Hook. f., Fl. Brit. India 1: 356. 1874; Malick in B.D. Sharma & Sanjappa, Fl. India 3: 468. 1993.

Shrub or small tree, rusty villous. Leaves simple, oblong or obovate-oblong, glabrous above, downy beneath; petiole 5–15 mm long. Flowers in many-flowered panicles; sparsely hispid; calyx deeply cleft. Male flowers: staminal column 2–3 mm long. Follicles 3–5 mm long, downy externally; seeds black.

India, Myanmar, and Malesia.

Fl.: December to February; Fr.: March to May.

Sterculia setigera Delile, Voy. Méroé 4(prepr.): 61. 1826.

Deciduous tree with a large, open, spreading crown; it generally grows up to 16 m tall, but specimens up to 35 m tall have been recorded in the Sudan and Guinea zones. Leaves are simple and alternate in arrangement, nerves palmately arranged at base with leaves 5 entire or pointed lobes; leaf-blade is ovate to orbicular in outline, upper surface tending to be stellate and tomentose. Inflorescence is raceme type, calyx 5-lobed, up to 12 mm in diameter; outer surface is green, while inner surface is purple-red. Fruit is a sessile follicle, 6–10 cm long, oblong in shape, grey-green or brown in colour, and many-seeded; follicle can stay on tree even when seeds fall off.

West tropical Africa – Senegal to Eritrea, south to Angola.

Fl. and Fr.: February to April.

Sterculia striatiflora Masters in Hook. f., Fl. Brit. India 1: 356. 1874; Baruah et al., Plant Sci. Today 8: 560. 2021.

Shrubs, 1–1.5 m high, bark greenish-grey. Leaves 15–25 cm × 5–8 cm, simple; lamina elliptic to oblanceolate, entire sometimes slightly undulate, caudate-acuminate, base cuneate, papery, only sparsely stellate-hairy above and thinly below mainly on nerves; petioles 2–3.5 cm, thickened just below the lamina; pulvinous terete. Stipules free lateral, narrowly subulate-acuminate, 4.5–6 × 1.0–1.2 mm, green, turns brown with age, herbaceous, stellate-hairy. Panicles axillary, 7–10 cm, male and female flowers on same inflorescence, lax, drooping, branches capillary. Pedicels short, 3–5 × 0.8–1 mm long, capillary. Bracts subulate, 3–3.5 × 1–1.5 mm, light-green, persistent; peduncle, rachis, petiole, and bracts densely hairy. Flowers dioecious, drooping, 8–10. Male flowers: staminal column 9.8–10 mm long, curved downwards, glabrous; anther lobes 5, somewhat triangular, yellow, dehiscing longitudinally; pollens yellow, granular. Female flowers: gynandrophore 6 mm long, broadened towards the tip and 5-lobed, each bearing one sterile anther; ovary 2 mm long, densely villous, hemispherical with 5 lobes; style 2.5 mm long, curved downward, greenish-white; stigma 5-lobed. Fruits not seen.

Myanmar (Burma), India (Arunachal Pradesh [East Siang District, Papum Pare District]; Assam [Dullung Reserve Forest, Lakhimpur]) (Borah and Das, 2018; Baruah et al., 2021).

Fl.: February.

Sterculia urens Roxb., Pl. Coromandel t. 24. 1795; Masters in Hook. f., Fl. Brit. India 1: 355. 1874; Gamble, Fl. Madras 1: 106. 1915; Malick in B.D. Sharma & Sanjappa, Fl. India 3: 470. 1993. *Kavalama urens* (Roxb.) Raf., Sylv. Tellur. 72. 1838.

Vern.: Hindi: Kulu, Gulu, Gular; Malayalam: Thondi; Marathi: Pandruk, Kandal, Karai; Santali: Keonji, Karaunji, Telhec; Tamil: Kavalam, Senthanakku, Senthalamaram; Telugu: Tabsu, Konda tamara, Kovila chettu, Thapsi chettu, Yerra poliki. English: gum karaya tree.

Soft-wooded, deciduous tree, up to 15 m high, with white papery outer bark, exfoliating in large thin irregular flakes, inner bark fibrous; trunk erect, branches spreading, marked with large scars, wood reddish-brown. Leaves palmately 3- or 5-lobed, crowded at the tips of the branchlets, 12–15 × 10–15 cm, glabrous above, velvety below, base cordate, margin entire, apex acuminate; petioles 8–9.5 cm long, terete; stipules caducous. Flowers greenish yellow in terminal panicles; panicles clothed with dense sticky tomentum of glandular stellate hairs; male and bisexual flowers are mixed. Male flowers: stamens 10. Female flowers: ovaries 2 mm in diameter on a ca 3 mm gynandrophore, with sterile anthers at base; styles as long as ovary, stigmas radiating. Follicles 5, oblong, radiating, yellow, pubescent, with stinging hairs; seeds 3–6, oblong, black, glossy **(Figures 2.1–2.6)**.

A common species in deciduous forests, on the dry, steep, rocky slopes in India, Sri Lanka, and Malaysia.

Fl.: December to March; Fr.: April onwards, dehiscing by June.

Chromosome number 2n = 40 (Raghavan and Arora, 1958).

Sterculia versicolor Wallich, Pl. Asiat. Rar. 1: 48, t. 59. 1830; Masters in Hook. f., Fl. Brit. India 1: 355. 1874; Malick in B.D.Sharma & Sanjappa, Fl. India 3: 472. 1993.

Vern.: Assamese: *Durong-phang;* Lus.: *Khai-phang-thing*; Kh.: *Star-um.*

Deciduous tree, up to 6 m high, trunk straight, bark grey, exfoliating. Leaves peltate, digitately 5- to 7-foliolate, petioles 15–45 cm long; leaflets elliptic-lanceolate,

FIGURE 2.1 *Sterculia urens* tree.

Source: Kothareddy Prasad, with permission.

FIGURE 2.2 *Sterculia urens* stem.

Source: Kothareddy Prasad, with permission.

FIGURE 2.3 *Sterculia urens* leaves and inflorescence.

Source: Kothareddy Prasad, with permission.

FIGURE 2.4 *Sterculia urens* flower.

Source: Kothareddy Prasad, with permission.

FIGURE 2.5 *Sterculia urens* immature fruits.

Source: Kothareddy Prasad, with permission.

FIGURE 2.6 *Sterculia urens* mature fruits.

Source: Kothareddy Prasad, with permission.

glabrous above, puberulous or glabrescent beneath. Flowers orange-yellow or pale yellow, in erect panicles crowded at the ends of branches; calyx campanulate, hairy. Male flowers: staminal column c. 2 mm long, curved. Female flowers: ovaries on 4–5 mm long gynandrophore, 5-lobed. Follicles 5, covered with dark-brown urticating bristles; seeds black, shining.

India and Myanmar.

Fl.: March to June

Sterculia villosa Roxb. ex DC., Prodr. 1: 483. 1824; Masters in Hook. f., Fl. Brit. India 1:355. 1874; Malick in B.D. Sharma & Sanjappa, Fl. India 3: 472. 1993.

Vern.: Assamese: *Udal, Odla*; Bengali and Hindi: *Udal*; Kannada: *Savaya, Bilidale*Malayalam: *Vakka*; Marathi: *Sardol*; Santali: *Ganghar*; Tamil: *Muruthan*; Telugu: *Gugal, Gogal, Kovila, Vakkunara, Kummari poliki.*

Tall deciduous trees, up to 20 m high, with white bark, branches few, horizontal, spreading with large scars. Leaves suborbicular, 30–45 cm each way, glabrescent or thinly stellate-pilose above, base cordate, deeply palmately 5- to 7-lobed, lobes acuminate. Flowers in drooping panicles, sepals pale-yellow and hairy outside, deep-red or purple within, lobed below middle. Male flowers: staminal column 2–3 mm long, curved, anthers 10. Female flowers: ovaries 5 on 2–3 mm long gynadrophore, globose with sterile anthers at base, style recurved, stigmas 5-lobed. Follicles bright reddish-purple, oblong, rusty villous, hairs not irritant, 7 × 2 cm; seeds oblong, black.

Occasional in dry rocky hill slopes of deciduous forests in India, Nepal, Bhutan, Bangladesh, and Myanmar.

Fl.: January to March; Fr.: May to June.

2.1.1 Chromosome Numbers

Base number *Sterculia* $n = 18, 20$

S. apetala $2n = 40$

Sterculia urens $n = 20$

2.1.2 Reproductive Biology

Sunnichan et al. (2004) investigated the reproductive biology of *Sterculia urens*. The species is andromonoecious and produces a large number of male and a limited number of 'bisexual' (functionally female) flowers. The numbers of male and 'bisexual' flowers vary not only between trees but also during the flowering period within a tree. Each male flower produces about 5,000 fertile pollen grains. Neither in morphology nor in number is there any difference between pollen grains in the 'bisexual' and male flowers. However, pollen grains of 'bisexual' flowers are completely sterile and incapable of siring any seeds. Their anthers, however, serve to attract pollinators; the emasculated 'bisexual' flowers fail to do so. Thus, *S. urens* is apparently andromonoecious but exhibits cryptic monoecy. That the species is self-incompatible was confirmed by controlled pollinations. The self-incompatibility is of the late-acting type and manifests after the entry of the pollen tube into the ovule. *Apis indica* is the only pollinator recorded by the investigators, and wind plays no role in pollination. The efficacy of pollination is low, as only 56% of flowers were estimated to be pollinated. The pollen load on one-third the number of pollinated stigmas was lower than the number of ovules present. Fruit set under open pollination is poor and is highly variable from tree to tree (0.7–3.2%). Apart from pollination constraint, limited resource availability may also contribute to low fruit set.

REFERENCES

Baruah, M., D. Dey, S.K. Boruah, M. Paul, B. Bhuyan and N. Devi. 2021. *Sterculia striatiflora* Mast. (Malvaceae s.l.)—A new addition to the flora of Assam. *Plant Sci. Today* [Internet]. 8(3): 560–563. www.horizonepublishing.com/journals/index.php/PST/article/view/1209

Borah D. and A.P. Das. 2018. Rediscovery of *Sterculia striatiflora* Mast. (Malvaceae) and its new record for the flora of India. *Pleione* 12(1): 105–109.

Gamble, J.S. 1915. *Flora of the Presidency of Madras*. Newman & Adlard, London.

Hooker, J.D. 1874. *Flora of British India,* Vol. 1. L. Reeve & Co., London.

Malick, K.C. 1993. *Sterculia* L. In: B.D. Sharma & Sanjappa, *Flora of India* (Bot. Survey of India), Vol. 3, pp. 457–472.

Raghavan, R.S. and C.M. Arora. 1958. Chromosome numbers in Indian Medicinal Plants – II. *Proc. Indian Acad Sci Section B*. 47: 352–358.

Sunnichan, V.G., H.Y. Mohan Ram and K.R. Shivanna. 2004. Floral sexuality and breeding system in gum karaya tree, *Sterculia urens*. *Plant Syst. Evol.* 244: 201–218. DOI: 10.1007/s00606-003-0095-x

3 Gum Karaya Production and Processing

Niranjan Prasad

3.1 INTRODUCTION

Some trees of the *Sterculia* species produce gum known as gum karaya (GK). The major sources of GK are *Sterculia urens*, *Sterculia setigera*, and *Sterculia villosa*. This chapter aims to document information on old and recent research done on gum karaya production and processing. Some of the major *Sterculia* species producing GK are listed in Table 3.1 along with their geographical distribution. *Sterculia urens* is also locally known by names like gulu, kadaya, kullo, India gum, ghost tree, etc. (Seaman, 1980). The gum produced from this tree is also called 'Indian tragacanth', which is of immense commercial importance. The properties of this gum are similar to those of gum tragacanth, but it is available at a lower cost compared with gum tragacanth. These reasons make *Sterculia urens* gum an ideal alternative to gum tragacanth (Alverson et al., 1999; Goldstein, 1954).

Sterculia urens trees grow in deciduous forests with crooked, short bole and peculiar white or greenish-grey, smooth bark, which, when shines in moonlight, gives a ghost-like appearance, due to which it is also sometimes called as ghost tree (Figure 3.1a). The bark is thick, powdery, exfoliating in thin, papery, large flakes with the reddish-colour blaze. Flowers of this tree are polygamous, greenish-yellow, in axillary panicles, with stinging hairs, and appears during December to March (Figure 3.1b). Flowers are bisexual, without petals, and hairy on both surfaces. Its fruit consists of five sessile, radiating, ovate-lanceolate hard, coriaceous carpels of about 7.5 cm long. The carpels are found covered on the surface with many hard bristles (Figure 3.1c). The fruits appear in the month of April to May (Thombare et al., 2018). In India, gum karaya production starts in November and continues up to June. Though gum production also takes place during the rainy season from July to October, it is not collected as the gum produced during these months is infected with mould and fungus and also contains lot of moisture.

Sterculia urens has origins in Asia. It is found in tropical regions of India and the dry forest of Sri Lanka and Myanmar. This tree is the main source of gum karaya (Nussinovitch, 2010; Sivaraj et al., 2017; Verbeken et al., 2003). In India, this tree is found in sub-Himalayan regions, Chhattisgarh, Rajasthan, Madhya Pradesh, Andhra Pradesh, Kerala, and Maharashtra (Silori et al., 2005; Sunnichan et al., 2004). It is found growing at altitudes 300–750 m in Northern and Central India (Galla and Dubasi, 2010; Roecklein and Leung, 1987; Anonymous, 1949). This tree is capable

DOI: 10.1201/9781003438960-3

TABLE 3.1
Distribution of potential gum karaya–producing *Sterculia* species across the world

Sterculia species	Distribution	References
Sterculia urens Roxb.	Native to the Indian subcontinent, including India, Pakistan, Sri Lanka, and Myanmar. Also found in Parts of Africa, including Senegal, Mali, and Sudan.	Nussinovitch, 2010
Sterculia villosa L.	Native to Asia. It is distributed in the Indian subcontinent in Bhutan, India, Nepal, and Pakistan.	Rai et al., 2020
Sterculia africana (Lour.) Fiori	*Sterculia africana* has been recorded in south-eastern Egypt, eastern Sudan, Ethiopia, Djibouti, northern Somalia, southern Tanzania, Mozambique, Malawi, eastern Zambia, Zimbabwe, Botswana, and Namibia's Caprivi Strip.	Hyde et al., 2020
Sterculia foetida L.	Indian subcontinent: India, Sri Lanka, Pakistan, Bangladesh, Myanmar, Australia, Djibouti, Eritrea, Ethiopia, Indonesia, Kenya, Malaysia, Oman, Philippines, Somalia, Tanzania, Thailand, Uganda, Yemen, Republic of Zanzibar, and also cultivated elsewhere in the tropics.	Silitonga et al., 2013
Sterculia apetala (Jacq.) H. Karst.	*Sterculia apetala* is a tree native to the Americas, from Mexico to Peru and Brazil. The species is present in North America, Central America, the Caribbean, and South America.	Acevedo-Rodriguez and Strong, 2012
Sterculia setigera Delile	Native to Africa, especially west tropical Africa, covering Senegal, Gambia, Guinea, Mali, Ghana, Togo, Benin, Nigeria, Cameroon, Central African Republic, Congo–Kinshasa, Sudan, Ethiopia, Eritrea, Angola.	Atakpama et al., 2014
Sterculia striata A. St. Hil. & Naudin	Native to almost all Brazilian regions.	Oliveira et al., 2000
Sterculia tragacantha Lindl	Native to tropical Africa: Sierra Leone to DR Congo, south to Angola, Zambia, and Tanzania.	Okon et al., 2019
Sterculia guttata Roxb. ex G. Don	Native to the Indian subcontinent.	Nussinovitch, 2010
Sterculia scaphigera Wall. ex G. Don	Indian subcontinent, Bangladesh, Myanmar, Malaysia.	Nussinovitch, 2010

FIGURE 3.1 Karaya (a) tree, (b) inflorescence, and (c) fruit.

FIGURE 3.1 Continued

of growing in areas of extreme temperatures (mean annual temperature of 10–40°C and less water availability (mean annual rainfall of 500 to 1,900 mm) (Rayadu, 2003; Roecklein and Leung, 1987).

In India, people use the *S. urens* tree for different purposes for their livelihood, which makes this tree prone to overexploitation. In Andhra Pradesh (India), 50% of the total of tribal people's income comes from non-timber plant produce (Reddy and Kumar, 2010). People in tribal areas in India exploit the maximum quantity of gum during summer, when other sources of livelihood get exhausted or are not available. In Rajasthan, this tree was in large numbers in the past, but now it has become rare due to overexploitation (Jain et al., 2005). In the production and export of karaya gum, India used to be the leader (Davidson, 1980). Due to over- and illegal exploitation and meagre distribution, this tree has been declared endangered in the hills of Aravalli (Kala, 2016; Nair et al., 1995; Nanadagopalan et al., 2015; Purohit and Dave, 1996; Reddy and Reddy, 2008).

The government of India has started an awareness and conservation programme to protect *S. urens* trees. Gum tapping is banned from this tree in Madhya Pradesh and some other states (Bhattacharya et al., 2003; Prasad and Bhatnagar, 1991). Scientists have started working on the development of a scientific method of gum tapping and a mass multiplication technique for this tree. Scientists have developed the gum inducer method of gum tapping using ethephon, which improves gum yield, reduces injury, and prevents damage to plants due to insects and microbes (Abib et al., 2013; Kuruwanshi et al., 2017; Nair, 2004; Vasishth and Guleria, 2017). Ethephon (2-chloroethylphosphonic acid) introduction in karaya tree triggers gummosis by inducing stress in the tree and developing gum cavities in sapwood which secret gum (Babu and Menon, 1989; Bhatt, 1987; Vasishth and Guleria, 2017).

Karaya gum is a unique natural product which is biodegradable and renewable. It finds application in food, pharmaceutical, and other industrial applications. Many book chapters and reviews are there on GK and their applications (Dhiman et al., 2019; Raj et al., 2021; Verbeken et al., 2003). This chapter is a compilation of information and knowledge on past as well as recent developments on GK production and processing.

3.2 PRODUCTION/TAPPING

Karaya gum is produced from *Sterculia* species of trees in response to biotic and abiotic stresses, namely, injury, drought, heat, high temperature, and infection. Natural production of GK takes place either from cracks developed on the tree trunk (Figure 3.2a–b) or, sometimes, from the root in termite-affected sites (Figure 3.2c). In India, GK is mainly produced from *Sterculia urens*. *Sterculia setigera* is another tree which significantly contributes in GK production in Senegal and Mali. A minor quantity of GK is produced from *Sterculia villosa* in Sudan, India, and Pakistan.

FIGURE 3.2 Natural gum exudation in karaya tree from (a) crack and (b) branch tip and (c) termite-affected root.

FIGURE 3.2 Continued

In India, commercial production of GK started in the 1960s from *Sterculia* species. In the beginning, gum producers followed the blazing method for GK production. Later, an improved method of GK tapping was developed and popularized in Andhra Pradesh by the Girijan Cooperative Corporation Ltd., Vishakhapatnam. The gum producers were trained for the application of sickle/blazing knife for making a blaze in place of an axe, a cleaner method of gum collection and handling, and an improved method of drying, cleaning, and storage (Coppen, 1995).

The best time for GK tapping is April to June in India, when temperatures remain high and humidity is low in the environment. The superior quality of GK is produced during this period (Vasishth and Guleria, 2017). In the rainy season, a major quantity of GK produced is washed away, and the remaining gum collected is of inferior quality, having low viscosity (Verbeken et al., 2003). The production of GK varies depending on locality, time of collection, age of the tree, trunk girth, number of blazes, etc. Production from a single tree may range from 0.5 to 4.5 kg per season.

The quality of GK depends on its colour and the foreign matter present in it. GK is classified into five grades, namely, hand-picked, superior no. 1 and 2, fair average, and shifting (Davidson, 1980; López-Franco et al., 2009). Different methods of GK tapping have been developed for sustainable gum production and the conservation of karaya trees, which are described in following sections.

3.2.1 Blazing Method

Superior plants respond to injury (natural or man-made) by producing resin/gum/gum-resin as a defence mechanism, which aids in the healing of the wound and guards it against microbial and fungal infection (Luján-Medina et al., 2013). By lighting fires or injecting chemicals into plants, humans take advantage of the phenomenon whereby plants produce resin, gum, and resin-gum. Both the plant's stem and its root can be used to produce these natural goods. One such tree, the karaya, yields gum that is important commercially and is mostly used in food and medicine. Different techniques for tapping and producing karaya gum have been developed and are now being used.

Through ducts found in the cortex and pith, certain plants respond to wounds by producing gum as a defence mechanism. According to Kuruwanshi et al. (2017), stem and root are the primary sources of karaya gum. Gum karaya, which has good adhesive characteristics, is made from the sap of the karaya tree (Upadhyay, 2017). Making a blaze, peeling the bark, and making a deep cut in the stem are methods used to tap gum artificially. The other method involves adding gum inducers to the tree. A summary of the several blazing techniques – conventional, rectangular, and semi-circular – that have been invented or are now in use is given.

Traditional blazing of a wound with an irregular shape. In the traditional blazing method, an axe is used to create a wound with an uneven shape on the stem of a tree (Figure 3.3a). Once the gum has dried, it is picked up by hand. Large lumps of gathered material are split up, and foreign objects like insects, leaves, and other things are manually picked out. These processes result in gum that is darker in colour and sells for a lesser price in both domestic and foreign markets (Bhattacharya et al., 2003).

Rectangular blazing. In traditional gum tapping techniques, the tree suffers damage from numerous wounds that are both larger in size and irregular in shape.

FIGURE 3.3 Different methods of blazing for karaya gum production: (a) irregular-shape blazing using axe, and (b) semi-circular blazing.

Sometimes the damage is so severe that the trees can no longer withstand it, which causes the death of these trees, which are valuable natural resources. Forest departments in Damoh and Jabalpur in the Indian state of Madhya Pradesh and the Forest Research Institute (FRI), Dehradun (India), adopted an improved method of gum tapping to overcome the drawbacks of the traditional method. According to this method's recommendations, tapping should be done on trees with girths greater than 900 mm, and the initial tapping blaze should be 150 mm tall, 100 mm wide, and 25 mm deep. For every 500 mm increase in girth over 900 mm, one blaze may be added. It was advised to gather gum at intervals of two weeks for 16 times during the tapping season. A 2 cm high bark above the preceding one should be removed after each harvest to keep the blaze fresh and promote more gum production. Following these suggestions throughout each season while ensuring adequate distance between nearby blazes, a six-year rest interval is offered before returning to the preceding blaze's location. By using this technique, a yield of 1.5 to 4.5 kg of gum per tree per year was obtained.

Semi-circular blazing. The Kovel Foundation, Vishakhapatnam, Andhra Pradesh (India), invented the semi-circular blazing method of gum karaya tapping (Figure 3.3b) for continuous and sustainable tapping of high-quality gum. A sharp sickle is employed in this technique to create a semi-circular blaze. For gum collection, a 15 cm × 15 cm square piece of transparent polythene sheet is pinned to the blaze with thorns. This keeps gum away from coming into contact with the bark after it has exuded and with human hands while being collected. The bamboo basket is advised in this approach for gathering gum from trees, and forceps should be used to remove foreign objects from gum while cleaning and grading. This approach also suggests using a higher platform made of a wooden pole, bamboo mat, and the polythene sheet for drying gum.

In the blaze method of gum tapping, ducts capable of producing gum are exposed due to blazing and produce gum. As a result, while bearing in mind the morphology and physiology of the tree, it is necessary to optimize the size, shape, and depth of the blaze based on the minimization of blaze area and maximization of gum-producing ducts exposed due to blazing.

3.2.2 Gum Inducer Method

Apart from the blazing method of gum tapping by making wounds in the tree, gum production also takes place in karaya trees when some gum inducer like plant growth regulator or inorganic acid is introduced into the tree system. The commonly used plant growth regulator is ethephon, while sulphuric acid is used as an inorganic acid. Both of these materials help in enhancing gum production by inducing stress conditions in the tree. Several researchers have studied the response of different gum inducers for karaya gum production as described later.

In order to determine how ethephon affected gum yield and blaze healing, Nair et al. (1995) conducted trials. Each tree had ten holes drilled around the stem, each measuring 5 mm in diameter and 20 to 35 mm deep. The holes were drilled at a height of 1 m above the ground. In order to prevent the inducer's backflow, the inward slope was provided with perforations. The other five holes (across from the treated holes) were kept as distilled water controls, while the first five holes in a row received ethephon treatment. In each hole, either 1 ml of distilled water or 1 ml of an ethephon solution containing 190/285/390 mg of active component per ml was inserted. In comparison to the control, the gum yield increased 20 times after the ethephon treatment. Seven trees treated with ethephon generated about 1.5 kg of high-quality gum. There was a variation in yield among the trees, which was probably caused by heterozygosity. After 30 days, the control group had stopped producing gum, but the treatment group had continued to do so. In ethephon-treated trees, a histological study revealed that extensive gum ducts or cavities had developed in the secondary phloem and wood, encircling each hole in a radius of 20 to 30 mm. After 45 days, thick tissue formed at the site of the hole and replaced the tissues that were destroyed (Nair, 2003).

The centre of ICAR-Network Project on Harvesting, Processing, and Value Addition of Natural Resins and Gums at the Indira Gandhi Agriculture University in Raipur conducted experiments and injected ethephon between May and June 2010 into ten gum karaya trees in the Kurrabhata village of Gariaband (Chhattisgarh). Using a hand drill or battery-operated drill, four holes measuring 6 mm in diameter and 50 mm in depth were drilled into each tree (Figure 3.4a–b). A syringe was used to inject 1 ml of ethephon into each hole (Figure 3.4c), while 1 cc of distilled water was injected as the control. With the aid of a branch that had been cut to the right size, the hole was filled. The gum was produced from the same hole that the gum inducer had been put into (Figure 3.4d). Per ethephon-treated tree, the average yield of gum was 320 g, whereas little gum was produced in control trees. A tree with a girth of 500 cm produced the most gum, at 650 g, and a tree with a girth of 76 cm produced the least

FIGURE 3.4 Gum inducer method of karaya gum tapping: (a) use of hand drill for making a hole for injecting gum inducer, (b) use of battery-operated drill for making a hole, (c) gum inducer injection using syringe, and (d) gum exudation.

FIGURE 3.4 Continued

FIGURE 3.4 Continued

gum, at 10 g. This gum tapping technique caused the least damage to the tree (*Annual Report, 2011–12, ICAR-Indian Institute of Natural Resins and Gums, Ranchi, India*).

In a 2017 study, Vasishth and Guleria used ethephon to tap gum karaya. For two years, various gum tapping methods (borehole vs. V blaze) were put to the test in an effort to standardize the process (Vasishth and Guleria, 2017). The drilling approach produced more gum than other methods. Sulphuric acid was shown to be less effective at inducing gum than ethephon. The highest gum yield (135.2 g) was obtained using a drilling technique that used an ethephon application rate of 4 ml at a concentration of 300 mg/ml in a hole that was between 15 and 20 mm in diameter and 50 mm in depth. The health of the tree was unaffected by the application of ethephon, and the peak period for gum production was from April to June.

Kuruwanshi et al. (2017) investigated a number of gum inducer techniques for tapping gum karaya as part of the ICAR Network Project on Harvesting Processing and Value Addition of Natural Resins and Gums. In the gum inducer tapping method, injections of ethephon (2.34, 3.12, and 3.9%) and indole-3-acetic acid (IAA) (400 and 800 ppm) were given into the hole created by a battery-operated drill that was 6 mm in diameter and 25 mm deep (Figure 3.4b). Each tree had two holes drilled onto it, and a 2 ml chemical injection was given using a syringe (Figure 3.4c). Maximum gum production (0.717 and 0.777 kg) was achieved by applying 2 ml of ethephon, at a concentration of 3.9%, in two consecutive administrations at intervals of 45–60 days from April to June on a tree. In the investigation, it was found the gum was barely soluble in water, but it became more soluble in hot water. In an organic solution, it was found to be insoluble. It had a light acidity.

Researchers have explored a variety of chemical gum inducers. However, natural gum inducer has not been investigated. There is scope to investigate naturally occurring gum inducers, like the various acids found in citrus fruits, chillies, cashew fruit, etc. The gum produced by the gum inducer method also needs to be studied for the presence of chemical gum inducer residue, as most of the gum is produced from the same hole that the gum inducer is injected into.

3.2.3 Mechanization of Gum Karaya Production/Tapping

The employment of an axe (Figure 3.3a) and a blazing knife for creating cuts (Figure 3.3b) in a blazing method of gum karaya tapping was reported by Prasad et al. (2012) in their overview of work done for the mechanization of gum and resin tapping. They also reported employing a hand drill (Figure 3.4a) and a battery-operated drill (Figure 3.4b) to drill holes in the gum tapping technique using various gum inducers. A blazer (Figure 3.5a–b) was developed at ICAR-Indian Institute of

FIGURE 3.5 Blazer for making blaze on karaya tree for gum tapping: (a) blazer and (b) blaze made using a blazer.

FIGURE 3.5 Continued

Natural Resins and Gums, Ranchi in 2019 to make blazes on the karaya gum tree for gum exudation. For sustainable gum production and gum tree conservation, the usage of the blazer for blaze-making is important, which regulates the size and depth of the blaze.

The gum inducer approach leads to the producing gum of higher quality than traditional tapping techniques (Krishnan et al., 2013; Sharma and Prasad, 2013). This might be due to the production of gum in the gum inducer method having a lighter colour and having less contaminants, such as bark and other foreign materials.

3.3 HARVESTING/COLLECTION, PROCESSING, AND STORAGE

3.3.1 Harvesting/Collection

As soon as the tree is blazed, gum karaya production begins. On the first day, the most gum is produced. In a traditional method of collecting gum, the gum formed on the blaze is allowed to dry and then collected in the shape of tears, either with the aid of a sickle or directly by hand. However, a polythene sheet is pinned beneath the blaze in the technique developed by the Kovel Foundation, Vishakhapatnam, for the collecting of gum. A new polythene sheet is placed for consecutive gum collection without touching the gum, and the gum gathered on the first sheet is removed along with it. In India, mostly gum tapping starts in November and continues up to June. The gum is also tapped in the months of September and October; however, the gum produced in these months possesses low viscosity.

3.3.2 Processing

Gum karaya that has just been harvested could have 20% (d.b.) moisture. Additionally, it has contaminants like sand, silt, bark, and leaves. Gum karaya is whitish during the time of exudation, but as soon as it touches bark or other foreign objects, it turns coloured. It is advised to collect exuding gum on a plastic sheet to prevent colour change. Although gum is produced during the rainy season, it is not collected since it is of very poor quality and frequently has fungal and moth infestations. A mature karaya tree typically yields 1 to 5 kg of gum every season.

Production and collecting of karaya gum take place in Senegal from September to January and from March to July. The collected gum is manually or mechanically processed and separated into two grades, hand-picked and standard. Comparatively speaking, Senegalese karaya gum is of lower quality than Indian karaya gum. In India, karaya gum is sold in five grades based on the colour and impurities it contains: hand-picked, superior no. 1, superior no. 2, fair average, and sifting (Davidson, 1980; López-Franco et al., 2009). The Bureau of Indian Standard specification IS: 12408–1988, which is primarily based on ash content, heavy metal content, water absorption, presence of *E. coli* and *Salmonella*, etc., describes the specifications of gum karaya food grade ('IS 12408: 1988 Specification for Gum Karaya, Food Grade', 1988). The quality of gum has improved over time as a result of greater understanding among Indian gum pickers for better methods of gum tapping, collection, processing, and storage.

Drying. About 30–35% of moisture is present in gum karaya at the moment of exudation. After it is exuded, gum naturally dries on the tree until it is collected. Approximately 20% of newly gathered gum is moisture. Depending on the climate, the collected gum is sun-dried for 5 to 15 days to reduce the moisture content to 10%. Additionally, sun-drying aids in gum quality improvement by whitening the gum's colour (Whistler, 1993). To prevent pets and kids from interfering with the drying process, the Kovel Foundation in Vishakhapatnam advised sun-drying gum karaya on a raised platform constructed of wooden poles, bamboo mat, and polythene sheet (Figure 3.6a). At ICAR-IINRG, Ranchi, a solar dryer (Figure 3.6b) was also used to dry gum karaya in order to speed up the drying process and prevent the gum from mingling with foreign objects while it dried, and encouraging results were found (*Annual Report, 2012–13, ICAR-IINRG, Ranchi, India*).

The most crucial primary processing unit operation in the processing of karaya gum is drying, which is second only to cleaning and grading. For the purpose of identifying appropriate drying processes and dryers, several studies have examined the drying kinetics and properties of karaya gum. Using a cabinet-style convective dryer, Singh and Singh (2014) investigated the drying kinetics of three classes of karaya gum (grades I, II, and III). The Page model for the prediction of drying kinetics best fit the experimental data from the investigation. According to the study, as air temperature rose, the drying process and drying rate accelerated. Additionally, it was noted that grade I had the lowest drying rate constant for a given temperature, and grade III had the highest drying rate constant.

FIGURE 3.6 Karaya gum drying (a) on a raised platform in the sun and (b) in a solar drier.

Using a hot air drier, Sahu et al. (2017) investigated the drying properties and kinetics of gum karaya grits with geometric mean diameters ranging from 4.5 to 99 mm. A constant airflow rate of 1.5 m/s was used throughout the investigation for five levels of temperature in the range of 40 to 80°C. The experimental data revealed that complete drying took place in the falling rate period, and the constant rate period was absent at all temperature levels. In contrast to other biological materials, the initial moisture removal rate was fast, followed by a slow decline before becoming very low. This was likely caused by the gum karaya's case hardening. Walde et al. (1997) examined the microwave drying and grinding properties of karaya gum using a domestic microwave oven and a domestic dry grinder, respectively. Research found that the microwave-dried karaya gum sample was crisp and required less energy to grind without losing any of its characteristics. This might be because microwave drying is more uniform, which eliminates the case hardening issue and requires less energy to grind than traditional sun-dried gum.

3.3.2.1 Cleaning and Grading

Cleaning and grading are carried out manually or mechanically in processing units in the majority of the nations that produce karaya gum. At the level of the gum pickers, very little primary processing is carried out. Gum karaya is cleaned by removing extraneous substances (bark, leaves, soil, sand, etc.) once it is received in the processing facility. While finer particles like soil and sand are removed by sifting, larger foreign matter like bark and leaves are removed by hand-picking or winnowing. Gum karaya is categorized into several grades based on its colour and impurities that are still present after the majority of foreign objects have been removed. In comparison to lower-grade gum, higher-quality gum is lighter in colour and has fewer impurities.

3.3.3 Storage

More than 10% moisture content (d.b.) in karaya gum makes it more susceptible to fungus attack and colour change from white to dark brown. The gum must therefore be dried before being stored in order to reduce the moisture content to less than 10% (d.b.). The market value of gum with a fungal infection and a dark-brown colour is low. The ability of karaya gum to create a viscous, colloidal solution is a key factor in its use in several industries. Because karaya gum contains acetyl groups, it has this characteristic. If the gum is kept in arable containers in a warm, low-humidity environment, it is susceptible to losing its acetyl groups. Additionally, in the case of powdered gum karaya, such loss is faster. The gum is kept in cold storage, where the temperature is kept below 25°C, in airtight containers, to prevent the loss of the acetyl group.

For quality preservation, studies were done to manage the de-acetylation of gum karaya during storage. The samples of gum karaya were stored in ambient and cold settings (a household refrigerator) and packaged in low-density polyethylene (LDPE), high-density polyethylene (HDPE), aluminium-coated plastic (ACP), and polypropylene (PP) sheets with air and nitrogen filling. For one year, stored samples were examined for quality characteristics, such as loss on drying, swelling, volatile acid, and water absorption. Gum samples stored in an HDPE pouch with nitrogen filling and held in cold temperatures showed the highest gum swelling index.

3.4 CONCLUSIONS

Information on the production and processing of karaya gum is provided in this chapter. This gum is deemed safe for use in culinary, pharmaceutical, and industrial applications due to its biodegradability, biocompatibility, and renewability in the living system, as well as its affordability, ease of handling, and cost-effectiveness. The distinctive qualities of karaya gum give it an advantage over its competitors. The literature reviewed also indicates a decline in the production of gum karaya as a result of overexploitation, which has led to a decline in the number of karaya trees. In order to produce karaya gum sustainably, it is necessary to adopt a scientific method of gum tapping and expand the resource base. There is growing pressure on this tree to fill the supply–demand gap because it is only found in the wild and has hardly ever been intentionally planted. The necessity and potential for doing research to boost karaya gum tree yield are enormous.

REFERENCES

Abib, C.F., M. Ntoupka, R. Peltier, J.M. Harmand and P. Thaler. 2013. Ethephon: A tool to boost gum arabic production from *Acacia senegal* and to enhance gummosis processes. *Agrofor. Syst.* 87: 427–438. https://doi.org/10.1007/s10457-012-9564-y

Acevedo-Rodriguez, P. and M.T. Strong (Eds.). 2012. Catalogue of seed plants of the West Indies. Smithsonian Institution Scholarly Press, Washington, DC, pp. 1–1192.

Alverson, W.S., B.A. Whitlock, R. Nyffeler, C. Bayer, D.A. Baum, W.S. Alverson, B.A. Whitlock, R. Nyffeler, C. Bayer and D.A. Baum. 1999. Phylogeny of the core Malvales: Evidence from ndhF sequence data. *Am. J. Bot.* 86: 1474–1486.

Annual Report. 2011–12. ICAR-IINRG, Ranchi, India.

Annual Report. 2012–13. ICAR-Indian Institute of Natural Resins and Gums, Ranchi, India.

Anonymous. 1949. The wealth of India : A Dictionary of Indian raw materials and industrial products (Industrial Products—Part I). CSIR- National Institute of Science Communication and Information Resources, New Delhi.

Atakpama, W., F. Foléga, M. Dourma, Y. Woégan, B. Diwediga, K. Wala, K. Batawila and K. Akpagana. 2014. Woody species diversity, structure and distribution of *Sterculia setigera* Del. In Togo (West Africa). *Annu. Res. Rev. Biol.* 4: 4511–4528. https://doi.org/10.9734/arrb/2014/10732

Babu, A.M. and A.R.S. Menon. 1989. Ethephon induced gummosis in *Bombax ceiba* L. and *Sterculia urens* Roxb. *Indian For.* 115: 44–47.

Bhatt, J.R. 1987. Gum tapping in *Anogeissus latifolia* (Combretaceae) using ethephon. *Curr. Sci.* 56(18): 936–940.

Bhattacharya, P., B. Joshi and S.F. Hayat. 2003. An improved method of tapping gum from kullu *Sterculia urens*. *For. Trees Livelihoods* 13: 187–196. https://doi.org/10.1080/14728028.2003.9752454

Coppen, J.J.W. 1995. *Gums, resins and latexes of plant origin*. FAO Inter-Departmental Working Group, Rome.

Davidson, R.L. 1980. *Handbook of water-soluble gums and resins*. McGraw-Hill, New York.

Dhiman, M., S. Singh and M.M. Sharma. 2019. A review on *Sterculia urens* Roxb.: A boon to the livelihood for tribal people and industry. *Ind. Crops Prod.* 130: 341–351. https://doi.org/10.1016/j.indcrop.2018.12.065

Galla, N.R. and G.R. Dubasi. 2010. Chemical and functional characterization of Gum karaya (*Sterculia urens* L.) seed meal. *Food Hydrocoll.* 24: 479–485. https://doi.org/10.1016/j.foodhyd.2009.12.003

Goldstein, A.M. 1954. Chemistry, properties, and application of gum Karaya. *Nat. Plant Hydrocolloids* 33–37. https://doi.org/10.1021/ba-1954-0011.ch006

Hyde, M.A., B.T. Wursten and P. Ballings. 2020. *Flora of Zimbabwe*. [WWW Document]. URL www.cabdirect.org/cabdirect/abstract/20197200816 (accessed 8.2.22).

Jain, A., S.S. Katewa, P.K. Galav and P. Sharma. 2005. Medicinal plant diversity of Sitamata wildlife sanctuary, Rajasthan, India. *J. Ethnopharmacol.* 102: 143–157. https://doi.org/10.1016/j.jep.2005.05.047

Kala, C.P. 2016. Important gum yielding species *Anogeissus latifolia* (Roxb.) Bedd, *Boswellia serrata* Roxb. and *Sterculia urens* Roxb. : Ethnobotany, population density and management. *Appl. Ecol. Environ. Sci.* 4(3): 61–65. https://doi.org/10.12691/aees-4-3-2

Krishnan, P.R., J.C. Tewari and T.K. Bhati. 2013. CAZRI gum inducer for gum production from *Acacia senegal*: Potential and SWOT Analysis. *Ann. Arid Zone* 52: 39–44.

Kuruwanshi, V.B., P. Katiyar and S. Khan. 2017. Scientific approaches of gum tapping in gum karaya (*Sterculia urens* Roxb.) for high gum production. *Int. J. Curr. Microbiol. Appl. Sci.* 6: 3366–3374. https://doi.org/10.20546/ijcmas.2017.608.402

López-Franco, Y., I. Higuera-Ciapara, F.M. Goycoolea and W. Wang. 2009. Other exudates: Tragancanth, karaya, mesquite gum and larchwood arabinogalactans. In: Phillips, G.O. and P.A. Williams (eds.) *Handbook hydrocolloids*. Second Ed. Wood Head Publishing, New York, pp. 495–534.

Luján-Medina, G.A., J. Ventura, A.C. Lara-Ceniceros, J.A. Ascacio-Valdés, V.D. Boone-Villa and C.N. Aguilar. 2013. Karaya Gum: General topics and applications. *Macromol. An Indian J.* 9, 2013.

Nair, M.N.B. 2003. Gum tapping in *Sterculia urens* Roxb. (Sterculiaceae) using ethephon. In: *XII World forestry congress*. Food and Agricultural Organization, Quebec City, pp. 0148–B4.

Nair, M.N.B. 2004. Gum tapping in *Sterculia urens* Roxb. (Sterculiaceae) using ethephon. In: *United States Department of Agriculture Forest Service general technical report PNW, GTR-604*, pp. 69–74.

Nair, M.N.B., K.R. Shivanna and H.Y. Mohan Ram. 1995. Ethephon enhances karaya gum yield and wound healing response: A preliminary report. *Curr. Sci.* 69: 809–810.

Nanadagopalan, V., M.J. Gritto and A. Doss. 2015. GC-MS analysis of biomolecules on the leaves extract of *Sterculia urens* Roxb. *J. Pharmacogn. Phytochem.* 3: 193–196.

Nussinovitch, A. 2010. *Plant gum exudates of the world: Sources, distribution, properties, and applications*. CRC Press, New York.

Okon, J., M. Christopher, E.N. Okey and I.A. Ibanga. 2019. Nutraceutical potentials of wild and neglected edible leafy vegetables (*Sterculia tragacantha* and *Sesamum indicum*) in Akwa Ibom State-Nigeria. *Eur. J. Biomed. Pharm. Sci.* 6: 30–36.

Oliveira, J.T.A., I.M. Vasconcelos, L.C.N.M. Bezerra, S.B. Silveira, A.C.O. Monteiro and R.A. Moreira. 2000. Composition and nutritional properties of seeds from *Pachira aquatica* Aubl, *Sterculia striata* St Hil et Naud and *Terminalia catappa* Linn. *Food Chem.* 70: 185–191. https://doi.org/10.1016/S0308-8146(00)00076-5

Prasad, N., K. Sharma, J. Tiwari, R. Gupta and M. Thomas. 2012. Mechanization of natural gum and resin tapping. *J. Non-Timber For. Prod.* 19: 279–289.

Prasad, R. and P. Bhatnagar. 1991. *Socio-economic potential of minor forest produce in Madhya Pradesh*. Jabalpur. Mpsfri.org.

Purohit, S.D. and A. Dave. 1996. Micropropagation of *Sterculia urens* Roxb.—An endangered tree species. *Plant Cell Rep.* 15: 704–706. https://doi.org/10.1007/BF00231929

Raj, V., J.H. Lee, J.J. Shim and J. Lee. 2021. Recent findings and future directions of grafted gum karaya polysaccharides and their various applications: A review. *Carbohydr. Polym.* 258. https://doi.org/10.1016/j.carbpol.2021.117687

Rai, Y., P. Kumar, S. Rajput and P. Paliwal. 2020. Growth and development of rare tree species *Sterculia villosa* Linn. in districts, Meerut and Bulandshahr, UP, India. *Asian J. Res. Agric. For.* 6: 1–8. https://doi.org/10.9734/AJRAF/2020/v6i430109

Rayadu, V. 2003. Propagation and scientific collection of gum from *Sterculia* for sustainability. In: *XII World forestry congress*. Food and Agricultural Organization, Quebec City, pp. 0229–B2.

Reddy, K.N. and C.S. Reddy. 2008. First Red List of medicinal plants of Andhra Pradesh, India-Conservation Assessment and Management Planning. *Ethnobot. Leafl.* 103–107.

Reddy, M. and K. Kumar. 2010. *Political economy of tribal development: A case study of Andhra Pradesh*. Centre For Economic and Social Studies, Begumpet.

Roecklein, J. and P. Leung (eds.), 1987. *A profile of economic plants*. Transaction Publishers, New York.

Sahu, P., S. Patel, P.S. Pisalkar and P. Katiyar. 2017. Studies on drying characteristics of gum karaya (*Stericulia urens* Roxb.). *Trends in Biosciences* 10(3): 1114–1117.

Seaman, J. 1980. Guar gum. In: Davidson, R.L. (ed.), *Handbook of water-soluble gums and resins*. McGraw-Hill, New York, pp. 6; 1–19.

Sharma, S. and N. Prasad. 2013. Comparison of gum tapping techniques for gum Karaya—a gum of commercial importance. *J. Non-Timber For. Prod.* 20: 165–170.

Silitonga, A.S., H.C. Ong, H.H. Masjuki, T.M.I. Mahlia, W.T. Chong and T.F. Yusaf. 2013. Production of biodiesel from *Sterculia foetida* and its process optimization. *Fuel* 111: 478–484. https://doi.org/10.1016/J.FUEL.2013.03.051

Silori, C.S., M. Mehar, M.A. Khalid and V. Paul. 2005. Non-timber forest products: Conservation status and management priorities in the community managed forests of Andhra Pradesh, South India. *Int. J. Sustain. Dev. World Ecol.* 12, 334–346. https://doi.org/10.1080/13504500509469643

Singh, J. and B. Singh. 2014. Drying kinetics of gum karaya (*Sterculia urens*) using convective air dryer. *British J. Pharmaceut. Res.* 4(13): 1647–1656,

Sivaraj, N., S.R. Pandravada, K. Venkateswaran and N. Dikshit. 2017. Ethnic medicinal plant wealth of Eastern Ghats: Status, knowledge systems and conservation strategies. *Int. J. Curr. Res. Biosci. Plant Biol.* 4: 83–101. https://doi.org/10.20546/ijcrbp.2017.401.010

Sunnichan, V.G., H.Y. Mohan Ram and K.R. Shivanna. 2004. Floral sexuality and breeding system in gum karaya tree, *Sterculia urens*. *Plant Syst. Evol.* 244: 201–218. https://doi.org/10.1007/s00606-003-0095-x

Thombare, N., V. Lohot, N. Prasad and K. Sharma. 2018. *Major gum and resin plants of India: A field guide*. ICAR- Indian Institute of Natural Resins and Gums, Namkum, Ranchi.

Upadhyay, R.K. 2017. Nutritional, therapeutic, and pharmaceutical potential of plant gums: A review. *Int. J. Green Pharm.* 11: S30–S48.

Vasishth, A. and V. Guleria. 2017. Standardized gum tapping techniques to maximize yield from high-value Indian tree, *Sterculia urens*. *J. For. Res.* 28: 615–619. https://doi.org/10.1007/s11676-016-0315-1

Verbeken, D., S. Dierckx and K. Dewettinck. 2003. Exudate gums: Occurrence, production, and applications. *Appl. Microbiol. Biotechnol.* 63(1): 10–21. https://doi.org/10.1007/s00253-003-1354-z

Walde, S.G., K. Balaswamy, R. Shivaswamy, A. Chakkaravarthi and D.G. Rao. 1997. Microwave drying and grinding characteristics of gum karaya (*Sterculia urens*). *J. Food Eng.* 31: 305–313. https://doi.org/10.1016/S0260-8774(96)00082-9

Whistler, R.L. 1993. Exudate Gums. In: *Industrial gums: Polysaccharides and their derivatives:* Third Edition. Elsevier Inc., London, pp. 309–339. https://doi.org/10.1016/B978-0-08-092654-4.50016-4

4 Pharmacognosy of *Sterculia urens* and Allied Species

Lepakshi Md. Bhakshu and K. Venkata Ratnam

4.1 INTRODUCTION

Gum is an important exudate of trees which especially grow in tropical dry and moist deciduous forests that produce them in commercial scale. Gums have found an immense value in producing certain medicinal, industrial, and even food products. The gum exuded from the injuries on different parts of trees or shrubs is in the form of tears, nodulous, or amorphous semi-solids. This substance dries upon the contact with external air and sunlight and becomes hard, glassy clumps. Gum secretion may be enhanced by extreme temperatures and less moisture in the surrounding atmosphere. Gums are physiological by-products secreted from the injured or grafted tissues on mature parts of certain tree species (Sao, 2012). Among gum-secreting trees, the genus *Sterculia* plays an important role, as it possesses commercial value and is used in the production of diversified industrial products, apart from their medicinal applications. The present review focuses on the architectural studies of gum-synthesizing tissues to understand the scenario of gum exudation mechanism and the techniques developed for its enhanced production, based on literature reports, which might be helpful to people engaged in the conservation, collection, and processing of high-quality gum karaya. The gum karaya in India is represented by species such as *Sterculia urens* Roxb., which are a unique and popular non-wood forest products (kulo, Thapsi gum, gum karaya, kandol, karei, gulu, katilo, katiera katera as trade names), which needs urgent attention on the scientific aspects of gum secretion to minimize stress on the plant (Kuruwanshi and Pratibha, 2020; Mamta et al., 2019). The introduction of scientific methods of gum induction might relieve stress on the species created by the commercial demand for gum collection, which initially led the people to perform detrimental practices on the trees. Gum collection plays a vital role in the economy of several tribal people who directly have attachment with these trees, which requires maximum attention and needs training for safe collection and to enhance gum secretion rate. Such studies will also be helpful in the training programmes for primary gum collectors associated with the forests who also contribute to its sustainable utilization and the conservation of natural resources indirectly.

The Plant List includes 439 scientific plant names of species rank for the genus *Sterculia*. Of these, 60 are accepted species names (www.theplantlist.org/browse/A/

DOI: 10.1201/9781003438960-4

Malvaceae/Sterculia/). However, only a few of the *Sterculia* species have been evaluated in pharmacognostical or anatomical studies, and these are reviewed critically.

4.2 COMPARATIVE STUDY OF ANATOMICAL PROPERTIES OF *STERCULIA* SPECIES

Rao et al. (1996) made a comparative study on the structure of the vascular cambium of five *Sterculia* species – *S. colorata*, *S. alata*, *S. villosa*, *S. urens*, and *S. foetida*. It was discovered that cambium is arranged in a storied pattern, and cambial ray cells are isodiametric and axially elongated, or fusiform. Fusiform cambial cells in the tangential slice are hexagon-shaped, and cambial rays join with marginal sheath cells (Figure 4.1.F). The existence of multiple primary pit fields results in beaded

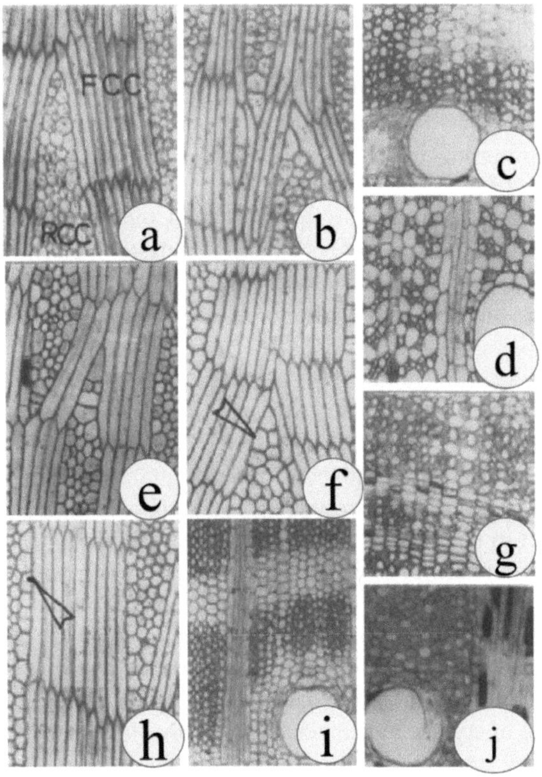

FIGURE 4.1 Anatomical studies of few *Sterculia* species: (a), (b), I, and (g), tangential view; (c), (d), (f), (h), and (i), transverse view. Storied arrangement of fusiform cambial cells in *Sterculia colorata* (4.1a), *S. alata* (4.1b), *S. villosa* (4.1e), *S. urens* (4.1f), and *S. foetida* (4.1g), × 75 (FCC = fusiform cambial cells; RCC = ray cambial cells; arrowhead = sheath cells). Distribution of axial parenchyma in *S. colorata* (4.1c), *S. villosa* (4.1d), *S. foetida* (4.1h), *S. alata* (4.1i), and *S. foetida* (4.1j); × 75.

Source: Rao et al. (1996).

radial walls in fusiform cambial cells. Radial longitudinal division produces two daughter cells of the same size in these cells. The cambial cells' general organization is nearly uniform, but each species' individual cell size, percentage, and relative quantity vary significantly. Cambial cells are in the size range of 302–424 µm on average. They are the longest in *S. foetida* (Figure 4.1h) and the shortest in *S. urens* (Figure 4.1f).

Of all the species examined, the most lengthy sieve tube members were found to be in *S. foetida*, measuring 449 µm, and the shortest in *S. urens*, measuring 291 µm (Figure 4.3). The members of the sieve tube ranged in width from 25 µm to 37 µmi in several species. On the other hand, *S. villosa* had the greatest width, measuring 37 µm, while *S. alata* had the lowest width, measuring 25 µm. The five species whose xylem was seen to have diffuse-porous growth rings were not very differentiated. However, vascular ray noding and the presence of small lumen cells with thick walls in the late wood could both indicate how much wood was created in a single development phase. Vascular rays would often nod in that area.

Vascular ray noding typically took place in the vicinity of the cambial zone in rest. Within each plant, the nodes vary in size.

The transverse section typically showed the nodes as spindle-shaped. The nodes might remain clearly in the forest. In certain trees, the nodding is less noticeable, though. The wood fibres had a small lumen and were non-septate. Five to eight times longer fibres than fusiform cambial cells were seen. *S. foetida* had the longest vessel element

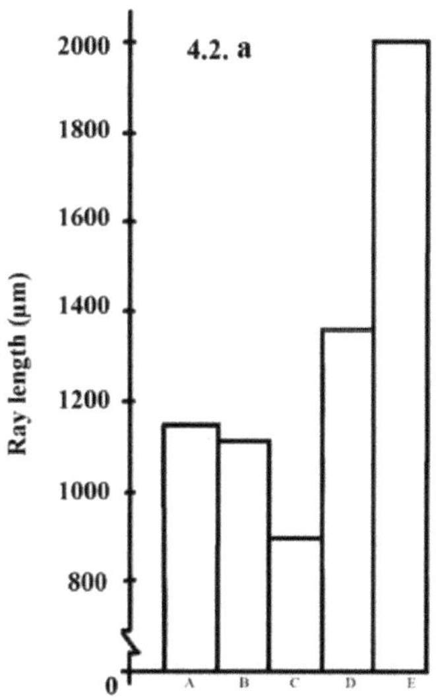

FIGURE 4.2a Histogram of mean cambial ray hight (µm) in (a) *S. colorata*, (b) *S. alata*, (c) *S. villosa*, (d) *S. urens*, and I *S. foetida*.

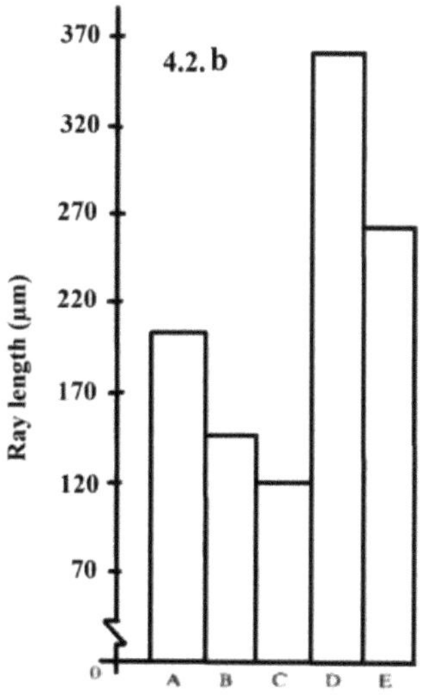

FIGURE 4.2b Histogram of mean cambial ray width (µm) in (a) *S. colorata*, (b) *S. alata*, (c) *S. villosa*, (d) *S. urens*, and I *S. foetida*.

Source: Rao et al. (1996).

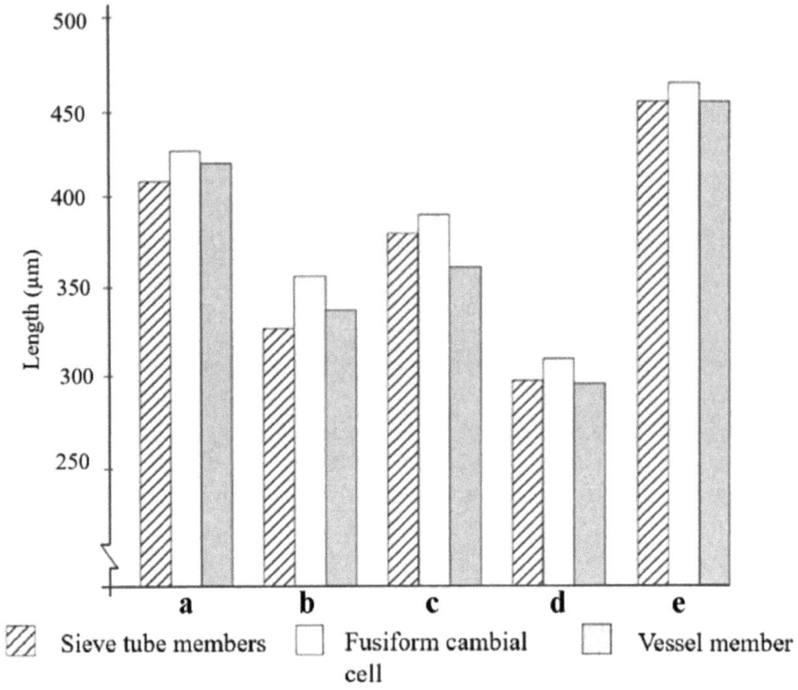

FIGURE 4.3 Histogram showing mean length (µm) of sieve tube elements, fusiform cambial cells, and vessel members in (a) *S. colorata*, (b) S. *alata*, (c) S. *villosa*, (d) S. *urens*, and I *S. foetida*.

Source: Rao et al. (1996).

length (450 µm), whereas *S. urens* had the shortest (285 µm) (Figure 4.2a,b). Vascular features were round to oval, solitary, diffuse, and in radial to tangential multiples of two or three in all five species. Their radial diameters ranged from 92 to 300 µm.

The vascular components' length had decreased compared to fusiform cambial cells (Figure 4.3 A-E), while their breadth had grown 6 to 12 times. Simple transverse to slightly oblique end wall perforation plates were used. The inter-vessel pits varied in shape from round to oval, with a diameter of 6 to 8 µm. In *S. colorata*, *S. alata*, and *S. urens*, the axial parenchyma was vasicentric to aliform, frequently forming distinct tangential confluent bands; in *S. villosa*, however, it was scarce paratracheal, and in *S. foetida*, it was scattered apotracheal. Phenolic deposits were present in some of the xylem ray cells in *S. villosa* and *S. foetida*. But in the other three species, no similar deposition was seen (Rao et al., 1996).

4.3 *STERCULIA FOETIDA*

The organoleptic study of seeds of *S. foetida* reported for the following characteristics: pale yellow colour, gritty feeling, bitter taste, and mildly aromatic scent. Important traits, such as typical tongue-sensitizing aromatic taste, aromatic odour, etc., are helpful in diagnosing different traits (Shaikh et al., 2017). The average ash values (%) for the three categories were 0.76 (acid-insoluble ash), 0.84 (water-soluble ash), and 3.9 (total). The comparatively low value of total ash could be attributed to the low concentration of silicates, carbonates, phosphates, and silica. The ash value is a valuable tool for assessing the legitimacy and purity of drugs, and it is also a crucial quantitative benchmark (Kokate et al., 2006).

The seed included 5.36% foreign organic materials; this could be the result of plant material that was collected directly from a non-polluted location; 7.65% of the weight was lost during the drying process, which is not too high and may prevent the formation of bacteria, fungi, or yeast. The lack of significant mucilage in the seeds may have contributed to the lack of noticeable swelling in the seeds after soaking. The value of crude fibre was determined to be 7.1%; determining the amount of crude fibre can help identify adulteration or discriminate between comparable medications. Additionally, it aids in the removal of plant organs' more resilient sections that can be utilized for microscopic inspection. Table 4.1 displays the mean values of the various solvent extractions.

Extractive value was highest in water and alcohol, indicating the possibility of considerable quantity of polar compounds in seeds. Preliminary phytochemical screening of alcoholic extracts indicated high concentration of flavonoids, alkaloids, along with other constituents (Table 4.1), like glycosides, saponins, and steroids.

It is known that these secondary metabolites have a variety of pharmacological effects, and it is possible that these effects are caused by the seeds of *S. foetida*. Table 4.3 presents the findings from fluorescence tests conducted on seed powder using various reagents, while Table 4.1 displays the results for extracts. The essential

TABLE 4.1
The phytochemical groups of secondary metabolites from successive extracts of *S. foetida* seeds

Constituents	Chloroform extract (CE)	Ethanol extract (EE)	Water extract (WE)
Alkaloids	+	+	–
Carbohydrates	–	+	+
Glycosides	+	+	–
Steroids	–	+	–
Flavonoids	–	+	+
Phenols	+	+	+
Saponins	+	+	+
Tannins	+	+	+
Proteins and amino acids	–	+	+

Source: Shamsundar and Paramjyothi (2010).

phenomenon of fluorescence is displayed by a variety of chemical components found in plant matter. In daylight, some components exhibit visible fluorescence. Many natural chemicals, such as alkaloids, like berberine, which do not fluoresce visually in daylight, fluoresce when exposed to ultraviolet light. Some crude pharmaceuticals are frequently evaluated qualitatively in this way because, even if the compounds themselves are not fluorescent, they may frequently be made into fluorescent derivatives by using different reagents. This is a crucial aspect of pharmacognostical evaluation (Shamsundar and Paramjyothi, 2010).

4.3.1 Macroscopical Studies

Organoleptic evaluation of stem bark (Figure 4.4a)

Colour:	Grey
Taste:	Bitter
Odour:	Odourless
Texture:	Smooth

Morphological evaluation:

Appearance:	Outer bark is smooth grey in colour; inner bark is fibrous and brown.
Shape:	Single quill
Length:	13–15 cm
Thickness:	2.5–3 cm
Fracture:	Fibrous

FIGURE 4.4a Morphology of the stem bark of *S. foetida*.

Source: Shamsundar and Paramjyothi (2010).

4.3.1.1 Organoleptic Study of Bark

The bark's transverse section (TS) of *S. foetida* showed different types of cells, such as cork cells, followed by cortex, cork cambium, sclerenchyma, endodermis, phloem cells, crystals of calcium oxalate, medullary rays, pericyclic fibres observed in different strengths of magnifications (Figure 4.4b).

Cork: It consists of 10–15 layers of regularly arranged rectangular cells in radial rows and with some reddish-brown content.

Cortex: It consists of parenchyma cells in with scattered sclereids are either isolated or in groups.

Medullary rays: It consists of parenchymatous cells dispersed diagonally and originating from pith or medulla to the cortex through secondary xylem and phloem, with a width of rays 4- to 5-celled multiseriate in a wavy pattern.

Phloem fibres: Phloem fibres are small, lignified, and polygonal.

Sclerenchymatous cells: The sclereids are lignified with pits. The inner and radial walls of the sclereids are thicker than the outer walls and appear in a U shape.

Calcium oxalate crystals: Prism-like crystals made of calcium oxalate were noticed in the cortical sclerenchyma.

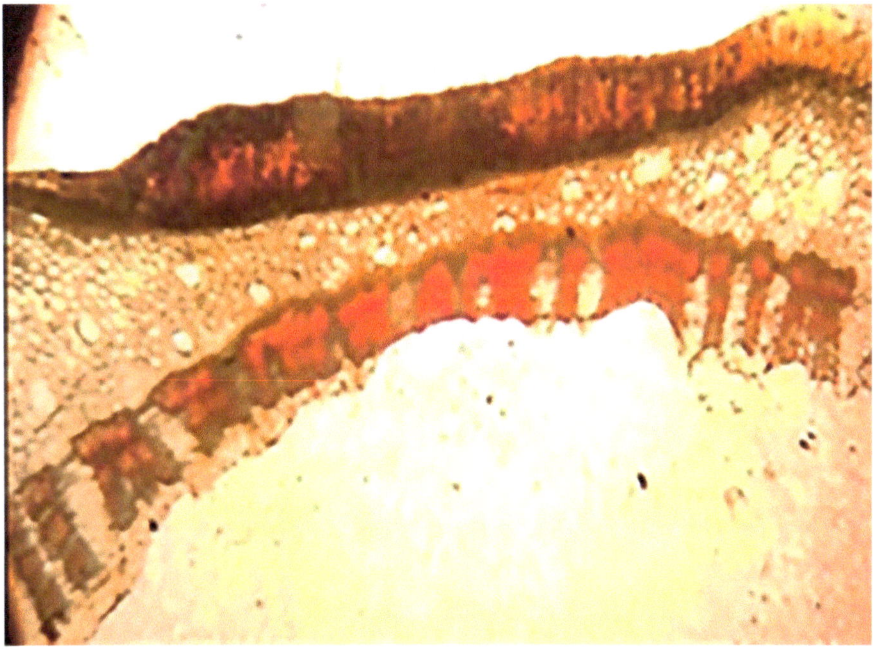

FIGURE 4.4b TS of bark of *S. foetida* showing different tissue layers arranged sequentially.

Source: Shamsundar and Paramjyothi (2010).

4.3.2 Histochemical Colour Reactions

Transverse section of the bark of *S. foetida*, when tested for the histochemical colour reactions of the transverse section of the bark (Figure 4.4c), showed different stained cells with specific features, as reported in Table 4.2.

The histochemical analysis on the bark of *Sterculia foetida* showed that the presence of alkaloids, starch grains, lignin, and flavonoids is localized in the cortex region; proteins and tannins are localized in the phloem region (Figure 4.5).

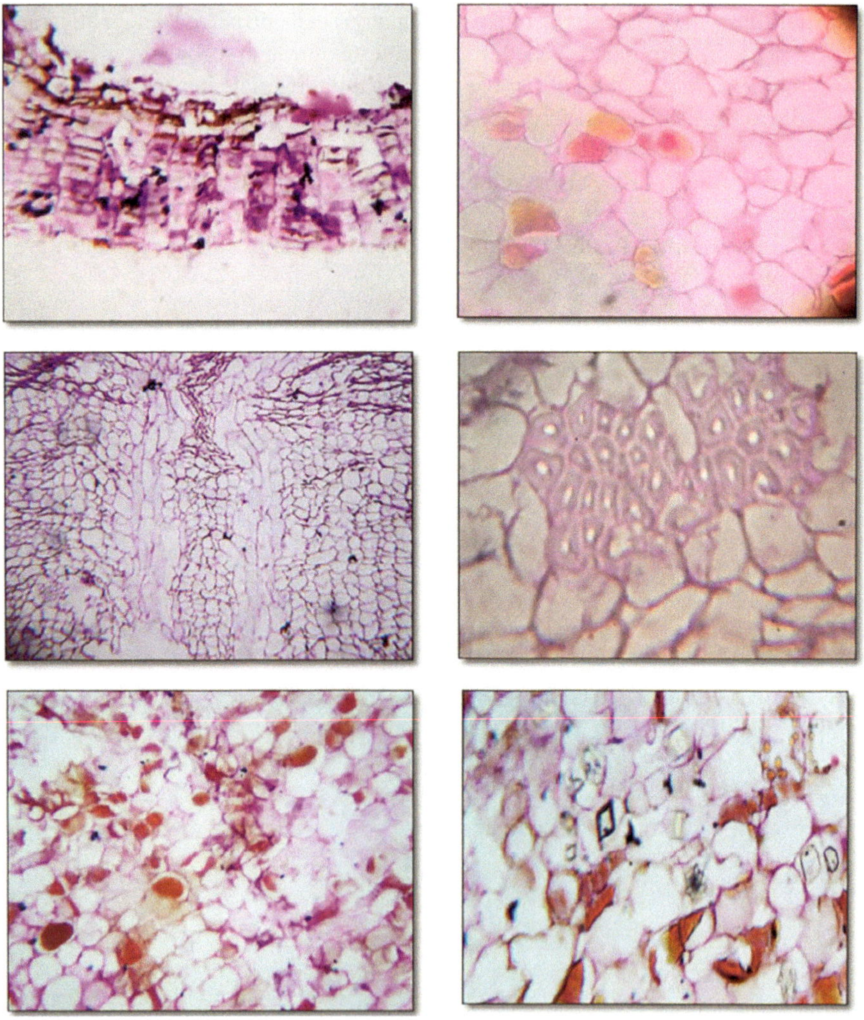

FIGURE 4.4c Histochemical staining indicates clear cellular architecture of stem bark of *S. foetida*.

Source: Shamsundar and Paramjyothi (2010).

TABLE 4.2
Histochemical colour reactions of *S. foetida*

S. no.	Chemicals	Test for	Nature of change	Histology	Degree of change
1	Phloroglucinol + HCl	Lignin	Dark yellow	In the cortex region	+
2	N/50 iodine solution	Starch	Black	In the cortex	+
3	Dil. ferric chloride	Tannin	Bluish green	Phloem parenchyma	+
4	Dragendorff reagent	Alkaloid	Brown	Outer cortex	+
5	Picric acid	Protein	Intense yellow	Phloem	+
6	Toluidine blue	Flavonoids	Bluish green	Cortex region	+

(*Source:* Shamsundar and Paramjyothi, 2010)

Note: +/– indicates presence and absence.

4.3.3 Histochemical Studies

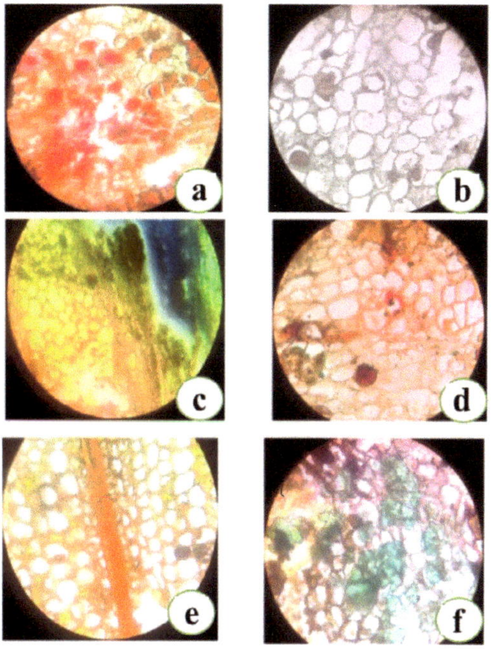

FIGURE 4.5a–e Localization of different secondary metabolites in the stem using staining with specific reagents: (a) pink indicating presence of lignin; (b) black indicating starch grains in the cortex; (c), blue-green indicates tannins in the phloem parenchyma; (d) brown colour indicates alkaloid in the outer cortex; (e) protein stained intense yellow in the phloem tissue; (f) bluish-green-stained locations indicate the presence of flavonoids in the cortex.

Source: Shamsundar and Paramjyothi (2010).

4.3.4 Powder Microscopy Organoleptic Characters (4.6a–e)

Nature:	Coarse powder
Colour:	Buff colour
Odour:	Characteristic odour
Taste:	Bitter taste

The microscopic study of powdered bark showed that *S. foetida* evidenced the presence of tissues such as xylem, lignified fibres, sclereids, phloem fibres crystals (calcium oxalate), and parenchyma (Shamsundar and Paramjyothi, 2010).

Fibres:	Lignified and scattered
Phloem fibres:	Groups of 3–5 cells attached with brown masses
Calcium oxalate crystals:	Prismatic and rosette type found in the cortical parenchyma
Cork:	Thin-walled cork cells, polygonal with brown mass
Sclereids (or) stone cells:	Ovoid to irregular and yellow (appear as 'U'-shaped differential thickness as inner wall is thicker)

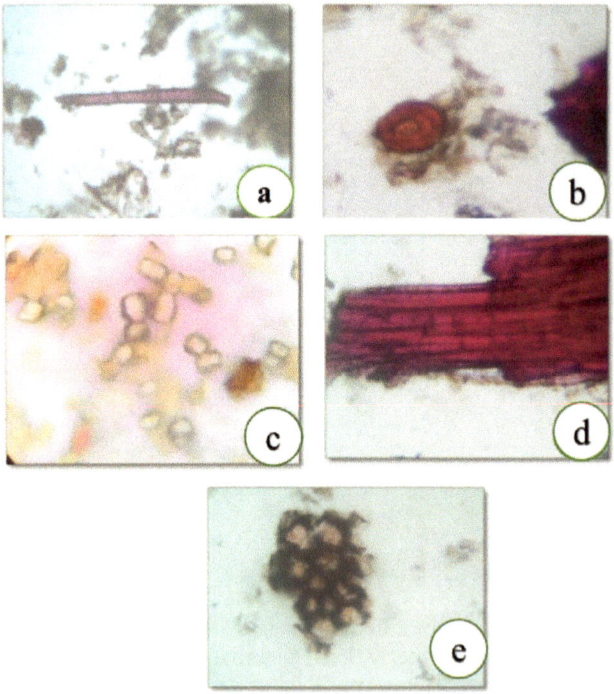

FIGURE 4.6 Tissue analysis in the macerate of stem bark of *S. foetida*: (a) fibres, (b) sclereids; (c) calcium oxalate crystals; (d) phloem fibres; (e) cork tissue.

Source: Shamsundar and Paramjyothi (2010).

4.3.5 Quantitative Microscopy

4.3.5.1 Linear Measurement of Fibres

The length and width of the fibres were measured in the powdered bark of *S. foetida*, and the metrics are shown in Table 4.3.

The length of the fibres was in the range of 365–423 µm, and the width of the fibres were found in the range of 49.8–54.78 µm (Table 4.3).

4.3.6 Physiochemical Constants

Physiochemical constants like total ash values, acid-insoluble ash, water-soluble ash, extractive values, loss on drying, swelling index, foaming index, and foreign organic matter are studied and reported in Table 4.4.

TABLE 4.3
The measurements of the fibres in the *S. foetida* bark

Dimension	Minimum (µm)	Average (µm)	Maximum (µm)
Length	365	423	531.2
Width	49.8	54.78	66.4

Source: Shamsundar and Paramjyothi (2010).

TABLE 4.4
Physicochemical properties of ash

S. no.	Physiochemical constants	Results (% w/w)
I	**Ash value**	
1.	Total ash	18.50 ± 0.16
2.	Water-soluble ash	8.26 ± 0.75
3.	Acid-insoluble ash	3.69 ± 0.33
4.	Sulphated ash	11.68 ± 0.36
II	**Extractive value**	
1.	Water-soluble extractive	3.56 ± 0.42
2.	Alcohol-soluble extractive	6.89 ± 0.53
3.	Ether-soluble extractive	5.63 ± 0.21
4.	Non-volatile ether-soluble extractive	4.64 ± 0.52
III	Loss on drying	1.88 ± 0.50
IV	Foaming index	Nil
V	Swelling index	Nil
VI	Foreign organic matter	0.96 ± 0.54

Source: Radha (2016).

Note: Values are expressed as mean ± SD; $n = 3$.

The organic matter of *S. foetida* bark was reported as 0.96 ± 0.54% w/w. The total ash, acid-insoluble ash, water-soluble ash, and sulphated ash contents were reported as 18.50 ± 0.16% w/w, 3.69 ± 0.33% w/w, 8.26 ± 0.75% w/w, and 11.68 ± 0.36% w/w, respectively. The water/alcohol/ether-soluble extractive and non-volatile extractive values were reported as 3.56 ± 0.42% w/w, 6.89 ± 0.53% w/w, 5.63 ± 0.21% w/w, and 4.64 ± 0.52% w/w, respectively (Table 4.4). Loss due to drying was reported as 1.88 ± 0.50% w/w. The foaming index and swelling index were reported as nil, which indicated the lack of mucilage and saponins (Radha, 2016).

4.4 *STERCULIA MACROPHYLLA* VENT.

Prastiwi et al. (2018a) have reported star-shaped trichomes as a specific fragment in the microscopic properties of leaves in brief as tissue system observed (Figure 4.7a–j).

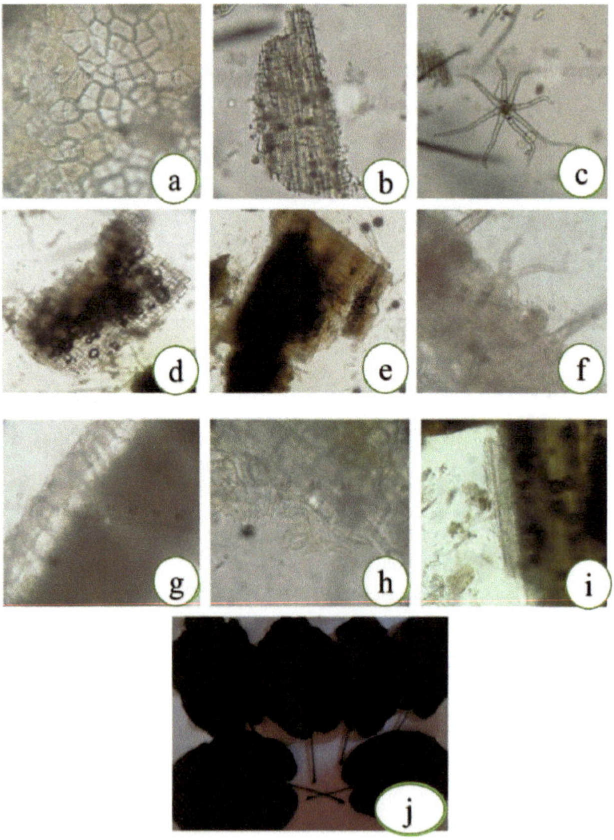

FIGURE 4.7 Microscopic and macroscopic views of *S. macrophylla* leaves powder: (a) parenchyma; (b) parenchyma; (c) star-shaped trichome; (d) parenchyma; (e) cortex; (f) star-shaped trichome on parenchyma; (g) epidermis; (h) stomata; (i) phloem; (j) leaves of *S. macrophylla.*

Source: Prastiwi et al. (2018a).

4.5 *STERCULIA QUADRIFIDA* R.BR.

S. quadrifida grows in Indonesia, India, Timor-Leste, and Australia. In East Nusa Tenggara province, Indonesia, particularly in Timor Island, *S. quadrifida* is called 'faloak', and its bark is used to treat hepatitis. Siswadi et al. (2021) determined the correlation of total flavonoids and bark colour, tree diameter class and altitudes, and examined the interaction between tree diameter class and altitudes in view of flavonoid content. *S. quadrifida* barks of four altitude strata of ≤300, >300–600, >600–900, and >900 m a.s.l. Each stratum of altitude consisted of three diameter classes of ≤15, >15–30, and >30 cm. Bark colour was classified using Munsell colour chart. The powdered barks were macerated with 96% ethanol and evaporated to yield thick extracts. There were four dominant colour groups, namely, red, brown, yellow, and orange. The highest-quantity flavonoid was reported in the red (0.25%) and yellow (0.25%) barks. The results showed that the highest total flavonoid content was found in tree diameter of >15–30 cm, with a total flavonoid of 0.22%. The highest total flavonoid (0.21%) was observed in the barks from an altitude range of ≤300 m a.s.l. Weak interaction is noticed between altitude to the tree diameter with the extract's yield and total flavonoid of *S. quadrifida* bark (Siswadi et al., 2021).

4.6 PHARMACOGNOSY OF *S. RUBIGINOSA*

It is a nutritious and traditional medicinal tree distributed in Indonesia and Borneo. The leaf powder was investigated for its microscopic properties, and star-shaped hair star recognition fragment, sclerenchyma fibres, oil cells, parenchyma, and anomocytic stomata type have been reported. Details are provided in Figure 4.8.

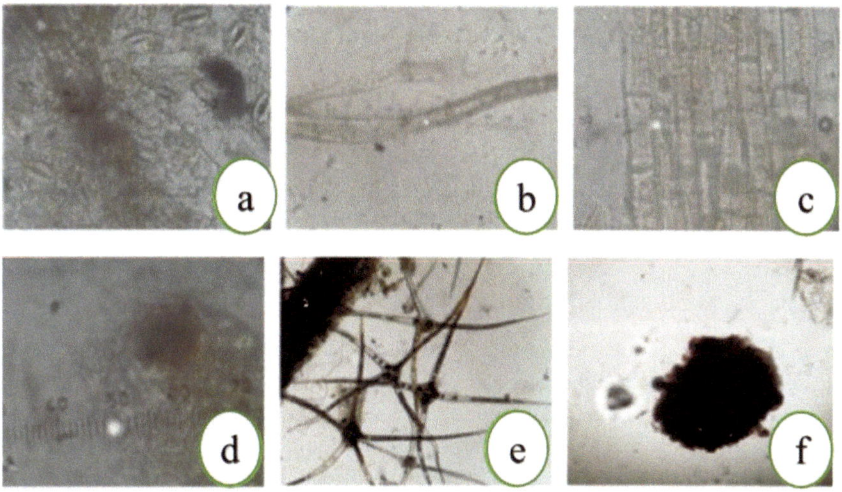

FIGURE 4.8a–f Diagnostic and microscopic features of leaf powder of *S. rubiginosa*: (a) leaf powder; (b) anomocytic stomata; (c) sclerenchymatous fibre; (d) parenchyma; (e) idioblast; (f) star-shaped trichomes oil cells.

Source: Prastiwi et al. (2018b).

The leaves showed star-shaped trichomes as a specific fragment. Physicochemical studies such as total ash (17.152%), acid-insoluble ash (0.922%), water-soluble extractive (1.610% w/w), alcohol-soluble extractive (4.5% w/w), hexane-soluble extractive (4.005% w/w), and ethyl acetate-soluble extractive (3.160% w/w) were evaluated. Phytochemical analysis of ethanol extracts reacted to the presence of tannins, flavonoids, alkaloids, steroids-terpenoids, glycosides, and phenols and the absence of saponins and anthraquinones. Antioxidant activity with IC_{50} 157.5 and flavonoid total was 59.5 mg/g quercetin equivalent (Prastiwi et al., 2018b).

4.7 *S. URENS* R.BR.

Babu and Menon (1989) explored the anatomical properties of *S. urens* and reported that the stem possesses normal gum ducts restricted to the cortex and pith, while they are absent in the wood (Figure 4.9a). Wood shows characterized banded parenchyma with broad multiseriate radial cells and thick-walled fibres. Extensive improvement is observed in the induction of gum cavities by injecting ethephon in the secondary xylem through the axial parenchyma (Figure 4.9b–c). Upon ethephon treatment, the ray cells remain intact, but only axial parenchyma cells are formed from the fusiform initials. The reactivated cambium produced the band of 'traumatic tissue' possessing ray as well as axial parenchyma in the peripheral sapwood (Figure 4.9d). The axial parenchyma divided actively and produced the isodiametric cells that vertically form cavity initials (Figure 4.9a) with the thin wall and dense cytoplasmic content with prominent nuclei (Figure 4.9b). Later, these groups of cells produce lysigenous cavities. The lysis of cells triggered during the disintegration of the vertical column of thin-walled cells precedes the disappearance of cytoplasm as well as nuclei. Progressive lysis of cells in vertical as well as tangential ways forms the cavities (Figure 4.9 B) without formation of epithelium. Widening of cavities in tangential direction does not affect the ray cells, which remain intact and find ray cells as islands (Figure 4.9 C). The gummy substance fills the cavities and extends by partial disintegration of ray cells in certain places (Figure 4.9 C-D). The cavities appear as longitudinally elongated, organizing the final gum-producing system interrupted by the ray cells arranged multiseriately (Figure 4.9 B) (Babu and Menon, 1989).

4.7.1 Gum Secretion Architecture of *S. urens*

In *S. urens*, the gum-exuding ducts are present in the pith as well as in the cortex of the young shoots (Shah and Setia, 1976), while they are absent in the secondary phloem. In artificial injury to the young as well as older stems in the course of 30–40 min, it is observed that there is fast recovery of disturbed ducts in the xylem during damage of xylem tissues, referred to as traumatic ducts (lysigenous nature). In addition, the presence of irregular lumen without traces of epithelium is also noticed (Figure 4.10a–d). Histochemical studies revealed the tendency of the gum secreted through the ducts to resemble that of normal ducts (Setia, 1984).

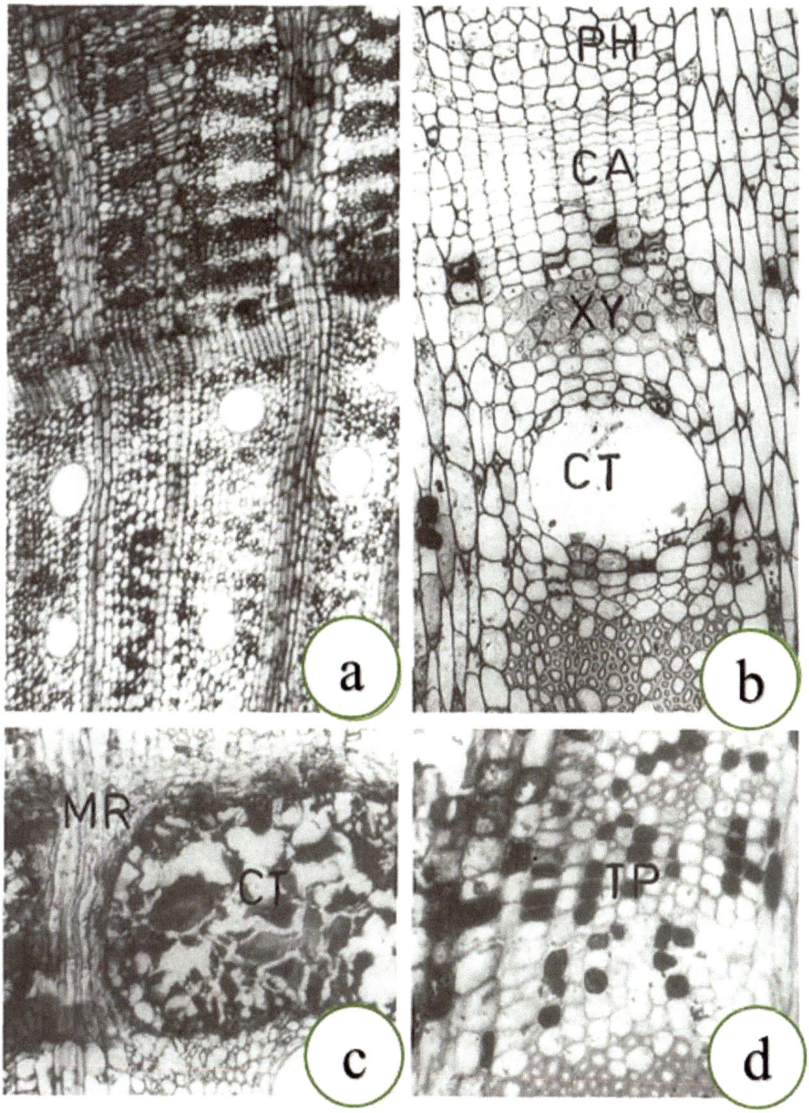

FIGURE 4.9 Transections from the outer secondary xylem of a stem: (a) normal structure of secondary xylem and phloem (note that the cavities are absent) (×50). Sections from ethephon-treated stem: (b) gum cavity in the secondary xylem near to cambium (×100); (c) a large cavity filled with gum (×90); (d) the parenchyma recovered in the outer (peripheral) sapwood during ethephon induction. **Note:** The vessels and fibres are absent in this tissue (×100). Where *CA*, cambium; *CT*, cavity; *MR*, multiseriate ray; *PH*, phloem; *TP*, traumatic parenchyma; *XY*, xylem.

Source: Babu and Menon (1989).

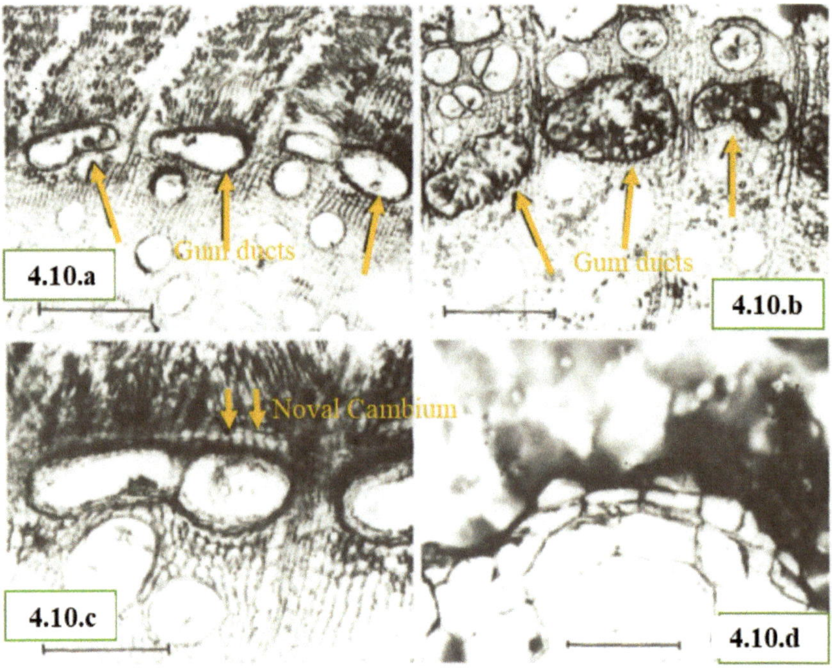

FIGURE 4.10a–d *S. urens* stem transverse section showing injury-induced formation of traumatic ducts. Arrows indicate the presence of traumatic ducts in the xylem in the vicinity of cambium. Arrows indicate the cells lining around the duct lumen, which do not contain gum.

Source: Setia (1984).

4.7.2 Mechanism of Duct Formation after Injuries/Wound

Observations on the transverse and longitudinal sections of the material showing gum exudation indicated the existence of tiny ducts near cambium in the xylem tissue (Figure 4.10a), referred to as 'traumatic ducts', since they are produced during injury or wounding. Similar ducts are also developed in the xylem parenchyma in the plant material fixed at 40 min interval after incision (Figure 4.10b). The traumatic gum ducts form a circular band parallel to the cambium zone (Figure 4.10c). A traumatic gum duct has irregular lumen, with no definite epithelial cell layer (Figure 4.10c), unlike a normal duct, which is oval to round in shape and has definite epithelial layer (Shah and Setia, 1976). The cambial zone is prominent and consists of 8–9 cell layers (Figure 4.10c). As this zone is structurally uniform irrespective of the absence or presence of the ducts nearby, it is concluded that the traumatic gum ducts develop from the young and mature xylem cells. The parenchyma cells lining the duct lumen show degradation and appear irregular (Figure 4.10d) and do not exhibit gum secretion, since the absence of a clear layer of epithelial cells is presumed to be of lysigenous nature in duct formation. The histochemical studies of normal/control as well as traumatic ducts indicated similar properties in young stems. Gum secretion in plants is a result of pathogen-induced metabolic processes or injury or as a regular metabolic activity (Setia and Shah, 1979). The case study of Setia (1984) opined that the injury induces the formation of traumatic gum ducts because of its exposure to the atmosphere of the external environment, which may be a chance for the invasion of microorganisms which induce pathological breakdown and formation of gum from the broken cells. This process might be helpful in inducing a high yield of gum from the trees by inducing mechanical tapping/injury on a large scale (Figure 4.10a–d).

4.8 METHODS TO ENHANCE GUM SECRETION

4.8.1 Gummosis Process in Stem

The secretion and tapping of gum are associated with the formation and architecture of the histological changes in the gum. The gum ducts normally appear in the pith and cortex of stems, while they are scanty or absent in the cork as well as the wood. Ethephon induces the extensive formation of gum cavities in the secondary xylem from the axial parenchyma.

While the axial parenchyma cells form from the fusiform initials, the ray cells are seen to remain intact. The cambium quickly resumes its regular activity, and as a result, the outer sapwood develops a zone of traumatized tissue made up solely of axial and ray parenchyma cells. Active transverse divisions occur in the axial parenchyma cells, and the derivatives grow to create vertical files of isodiametric cells, which are the cavity initials. They have big nuclei, thick cytoplasm, and usually thin walls. Initially, the cavity forms spontaneously from a collection of these cells. The cavities develop lysigenously during ethephon treatment.

The lysis is initiated by the disintegration of the cell's vertical file, which is followed by the cytoplasm becoming darker and the nuclei going missing. More cells lyse in tangential and vertical orientations, creating a hollow. But the area surrounding the cavities does not develop a distinct epithelium. The tangential enlargement of a hollow is restricted by largely intact multiseriate rays. Nearly all axial parenchyma cells lyse, creating a network of tangentially anastomosing voids around the multiseriate ray islands that remain intact. There are sticky materials and crumbling cells inside the cavity. Nevertheless, some multiseriate rays also break apart in areas where large cavities are formed. However, ray cells are always the last to suffer damage. The cavities appear as a vertically elongated system interrupted in radial longitudinal parts by multiseriate rays.

Shah and Setia (1976) noted that the juvenile stem of *S. urens* had lysigenously produced gum ducts in the pith and cortex but none in the xylem. When ethephon therapy is administered, gum cavities are created. According to Van der Molen et al. (1977), the cells whose cell walls are converted into gum may be either specialized parenchyma groups or mature xylem cells that develop in the cambium before disintegrating to produce the gum and duct lumen. Babu and Menon (1989), Setia (1984), and Wilde and Edgerton (1975) also published findings that were similar.

4.8.2 Rate of Gum Exudation in *S. urens*

Kuruwanshi and Pratibha (2020) examined the effects of gum production and how it was enhanced when ethephon was employed as an elicitor in their study on gum augmentation in *S. urens*. They discovered that ethephon had a great inducing impact, and anatomical evidence has experimentally supported this finding. The soft (sapwood) samples were taken for the biological (anatomical) investigations at various points in time after the gum enhancer ethephon was introduced. The results unequivocally showed that cavities that are capable of producing gum increase gummosis. In gum karaya, gum duct production was seen in the bark section histologically examined within two hours of ethephon treatment.

With the exception of the xylem, the histological alterations showed that the gum ducts lysigenously originated in the pith and cortex of the juvenile stem.

Using an axe or sickle, deep cuts are made at the base of the bole to accomplish the commercial tapping of karaya. These techniques frequently result in the death of the tapped trees and may induce population disappearance, which results in area extinction. These effects were attributed to overexploitation and shoddy tapping techniques, which resulted in a sharp fall in the number of karaya trees. The existing harvesting techniques are harmful and conventional, which frequently result in lower-quality crops.

As a result, Kuruwanshi and Pratibha (2020) enhanced the scientific tapping approach for sustainable *S. urens* harvesting in the state of Chhattisgarh by conducting a study in collaboration with ICAR. Kuruwanshi and Pratibha (2020) employed a battery-operated drill machine to inject ethephon and IAA at varying quantities in their chemical tapping approach. Nonetheless, cofactors like relative humidity and temperature also have a big impact on gummosis. In order to maximize gum production, ethephon 3.9% in 4 ml administered twice over 45–60 days at a high temperature in April through June was proven to be highly beneficial in both 2015 and 2016 studies.

According to Kuruwanshi and Pratibha's (2020) research, there is a notable fluctuation in the gum secretion rate from March to June. During the study period (2015 and 2016), May had the largest rate of exudation (440.33 and 530.1 g, respectively) compared to April and March, and June had the lowest rate of gum production. Compared to March and June, the month of May had the significantly greatest rate of gum exudation, followed by April. The reason for this could be that temperature and relative humidity were important co-inducers of gummosis and gum secretion. (Figure 4.11).

Figure 4.11 reveals that treatment of ethephon 3.9% (T4) is found to contribute to a significantly high yield (717.05 and 777.45 g), followed by treatments of 3.12% ethephon (T3) and 2.34% (T2). In the course of the experiment, treatment with IAA 800 ppm (T6) (21.43 g, 22.789 g) produced the lowest yield. However, the control (T1), which was treated with distilled water, produced a minimal amount of gum without any improvement in yield. Similar studies by Gupta et al. (2012) found that different doses of ethephon injected by drilling technique in the trunk of *S. urens* trees further enhanced gum exudation without affecting the health of the tree. Around 4 ml of ethephon injected twice, once in mid-March and once in the first week of May, per tree, also improved gum exudation.

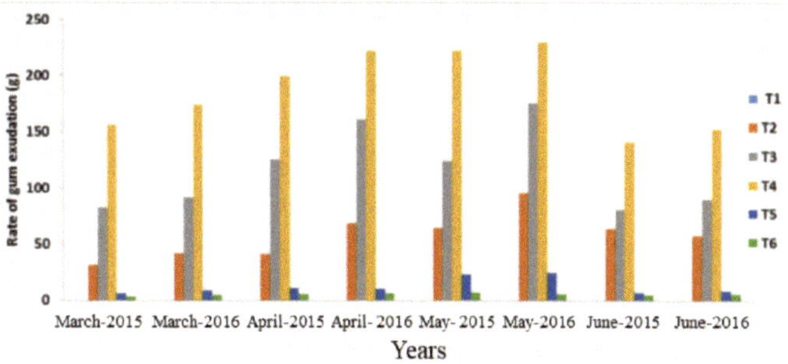

FIGURE 4.11 Effect of ethephon and IAA (as gum inducers) on rate of gum secretion (g) in gum karaya (*S. urens*) in the years 2015 and 2016.

Source: Kuruwanshi and Pratibha (2020).

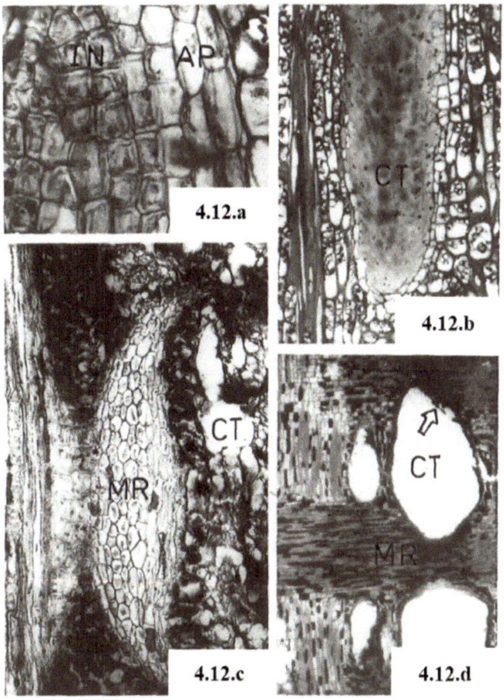

FIGURE 4.12 Stem section showing ethephon treatment: (a) tangential longitudinal section (TLS) showing vertical file of isodiametric cells (cavity initials) being formed by the transverse division of axial parenchyma cells; (b) radial longitudinal section (RLS) of a young cavity; (c) transverse longitudinal section (TLS) of an island of intact multiseriate ray in the gum cavity; (d) radial longitudinal section (RLS) showing two vertical rows of cavities with continuity of the cavities, which are interrupted by the multiseriate rays.

Source: Babu and Menon (1989).

Gum quantity was investigated by Babu and Menon (1989), who found that T4 with 3.9% ethephon produced the most gum (239.02 and 259.15 g), followed by T4 with 3.12% ethephon (137.51 g and 172.36 g), and T2 with 2.34% ethephon (66.80 g and 87.67 g). In contrast, T5 with 400 ppm IAA (15.84 g and 17.30 g) and T6 with 800 ppm of IAA produced the least amount of exudation when compared to treatment with 800 ppm (7.14 g and 7.59 g). But in T1 (control), 4 ml of distilled water did not improve the exudation that gum karaya plants were exhibiting. According to the pooled average results, the concentration of ethephon at 3.9% (249.08 g) had the greatest influence on the quantity of gum exudation, with concentrations of 3.12% (154.93 g) and 2.34% (77.23 g) following closely behind. More gum was formed than in the control when the gum enhancer ethephon was administered at concentrations of 3.9% and 3.12%. It might be because ethephon treatment caused cavities and gum ducts to develop in the *S. urens* stem (Babu and Menon, 1989). The investigations also provided an explanation using the treatments and architectural adjustments displayed in Figures 4.12a–d. Kuruwanshi and Prathiba's (2020) research demonstrated how anatomical alterations were correlated with subsequent treatment modifications, which unmistakably demonstrated the development of collagenous traumatized cavities following ethephon administration (Figure 4.13a–e).

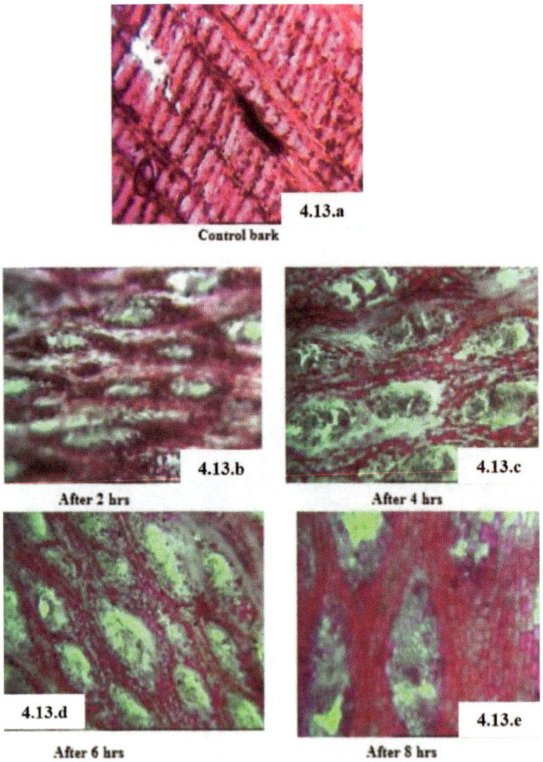

FIGURE 4.13a–e Anatomical properties of *Sterculia urens* Roxb. bark showing sequential formation of gum cavities induced by ethephon and gummosis process.

Source: Kuruwanshi and Pratibha (2020).

Gum yield increased as the temperature rose and the relative humidity dropped, and the two variables were positively correlated. When tapping was done during the abrupt drop in relative humidity, gum output was noticeably higher (Harmand et al., 2012).

Similar findings were also reported by Bhatt and Mohan Ram (1990), who found that by drilling a hole in the sapwood in April or May and injecting 4 ml of 'ethephon', or 960 mg of the active component, each tree could produce 0.8 to 0.9 kg of high-quality gum. The findings of Harsh et al. (2013) support it. In *Sterculia urens*, the administration of IAA resulted in an increase in gum duct count (Setia and Shah, 1977). Sharma and Prasad (2013) corroborated the idea that the application of ethephon causes sapwood, which clogs secondary xylem channels with gooey debris.

4.9 CONCLUSION

The detailed pharmacognostical investigations play a crucial role in the identification of plant drug in powdered form that is helpful in discrimination of adulteration. Gum karaya (from *Sterculia* spp.) became a rare commodity and was localized due to overexploitation for its gum. There is much appreciation on *S. urens* and less focus on other species. Its gum possesses enormous importance in applied aspects in various fields, such as pharma and allied science. Hence, to cope with the demands, scientific methods of tapping are required; more or less, gum collectors should be provided information on the tapping methods and get a license to harvest gum from only a few trees. Production of the gum and its enhancement are also equally important during gummosis for better yield and in reducing threat and risk on individual trees. Moreover, tribes are the main beneficiaries and get employment during non-agricultural periods. However, there is very little exposure on the gums of *Sterculia* species, except *S. foetida* and *S. urens*. Hence, further studies are needed by scientists for the betterment of human lives, besides the conservation aspects of the valuable commodity-yielding species.

REFERENCES

Babu, A.M. and A.R.S. Menon. 1989. Ethephon induced gummosis in *Bombax ceiba* L. and *Sterculia urens* Roxb. *Indian Forester* 115(1): 44–47.

Bhatt, J.R. and H.Y. Mohan Ram. 1990. Ethephon induced gum production in *Acacia senegal* and its potential value in the semi-arid regions of India. *Current Science* 59: 1247–1250.

Gupta, R., S. Patel, P. Katiyar and R.K. Modi. 2012. *Harvesting, processing and value addition of natural resin and gum*. Directorate of Research Services, IGKV, Raipur.

Harmand, J.M., M. Ntoupka, B. Mathieu, C.F. Njiti, J.M. Tapsou, J.C. Bois et al. 2012. Gum arabic production in *Acacia senegal* plantations in the Sudanian zone of Cameroon: Effects of climate, soil, tapping date and tree provenance. *Bois et forêts des tropiques* 311(1): 21–33.

Harsh, L.N., J.C.Tewari, H.A. Khan and M. Ram. 2013. Ethephon-induced gum arabic exudation technique and its sustainability in arid and semi-arid regions of India. *Forests, Trees and Livelihoods* 22(3): 204–211.

Kokate, C.K., A.P. Purohit and S.B. Gokhale. 2006. *Pharmacognosy*. 34th Edition. Nirali Prakashan, Pune

Kuruwanshi, V.B. and K. Pratibha. 2020. Anatomical study of ethylene induced gum duct formation and scientific gum tapping technique in gum Karaya (*Sterculia urens* Roxb.). *International Journal of Chemical Studies* 8(6): 2376– 2382. https://doi.org/10.22271/chemi.2020.v8.i6ah.11129

Mamta D., A. Singh, and M.M. Sharma. 2019. A review on *Sterculia urens* Roxb.: A boon to the livelihood for tribal people and industry. *Industrial Crops & Products* 130: 341 –351. https://doi.org/10.1016/j.indcrop.2018.12.065

Prastiwi, R., B. Elya, R. Sauriasari, M. Hanafi and Desmiaty, Y. 2018a. Arginase inhibitory, antioxidant activity and pharmacognosy study of *Sterculia macrophylla* Vent. leaves. *Pharmacognosy Journal* 10(6): 1109–1113.

Prastiwi, R., B. Elya, R. Sauriasari, M. Hanafi and E. Dewanti. 2018b. Pharmacognosy, phytochemical study and antioxidant activity of *Sterculia rubiginosa* Zoll. ex Miq. leaves. *Pharmacognosy Journal* 10(3): 571–575.

Radha, R. 2016. *Pharmacognostical, Phytochemical and Anti-Arthritic Activity on the Bark of Sterculia foetida Linn., and Isolation of Biological Marker*. A Dissertation submitted to The Tamilnadu Dr. M.G.R. Medical University, Chennai.

Rao, K.S., R.S. Kishore and T. Srinivas. 1996. Comparative structure of vascular cambium and its derivatives in some species of *Sterculia. IAWA Journal* 17(3): 311–318.

Sao, K.P. 2012. Physico-chemical properties of some Indian plant gums of commercial importance. Indian Institute of Natural Resins and Gums, Indian Council of Agricultural Research, Namkun, Ranchi (Jharkhand). *Technical Bulletin* 1: 1–64.

Setia, R.C. 1984. Traumatic duct formation in *Sterculia urens* Roxb. In response to injury. *Phyton* (Austria) 2: 253–255.

Setia, R.C. and J.J. Shah. 1977. Effect of morphactin IAA and kinetin on gum canals in *Sterculia urens* Roxb. *Indian Journal of Experimental Biology* 15: 297–301.

Setia, R.C. and J.J. Shah. 1979. Histological, histochemical and ultrastructural aspects of gum and gum-resin producing structures in plants. *Annual Review of Plant Science* 1: 315.

Shah, J.J. and R.C. Setia. 1976. Histological and histochemical changes during the development of gum canals in *Sterculia urens. Phytomorphology* 26: 151–158.

Shaikh, M.D., R.S. Dhotre and M.A. Kare. 2017. Pharmacognostical and phytochemical evaluation of *Sterculia foetida* L. stem bark. *International Journal of Scientific & Technology Research* 3(8): 1069–1072.

Shamsundar, S.G. and S. Paramjyothi. 2010. Preliminary pharmacognostical and phytochemical investigation on *Sterculia foetida* Linn. Seeds. *African Journal of Biotechnology* 9(834): 1987–1989.

Sharma, S.C. and N. Prasad. 2013. Comparison of gum tapping techniques for gum Karaya—a gum of commercial importance. *Journal of Non-Timber Forest Products* 20(3): 165–170.

Siswadi, S., E. Faridah and T. Hertiani. 2021. Total flavonoid content of faloak (*Sterculia quadrifida*) bark in varieties of bark colour, tree diameter and growth altitude. *Journal of Tropical Forest Science* 33(3): 298–307.

Van der Molen, G.E., C.H. Beckman and E. Rodehorst. 1977. Vascular gelation: A general response phenomenon following infection. *Physiological Plant Pathology* 11: 95–100.

Wilde, M.H. and L.J. Edgerton. 1975. Histology of ethephon injury on "Montamorency" Cherry branches. *HortScience* 10: 79–81.

5 Phytochemistry of Gum Karaya – *Sterculia urens* Roxb. and Its Allied Taxa

Lepakshi Md. Bhakshu and K. Venkata Ratnam

5.1 INTRODUCTION

Sterculia is a genus of about 60 species of trees mostly found in tropical and warm areas, but a few are found in temperate areas. It is a member of the flowering plant family Malvaceae. Though many species of the *Sterculia* are reported for medicinal uses, a few have been studied in detail for phytochemical characterization. *Sterculia urens* Roxb. is popularized as the 'gum karaya tree' spread around the world. In medicine, leaves are used to treat pleuropneumonia in cattle, as well as to trigger parturition and serve as a source of vitamin A. Its bark juice is used to cure piles, while its bark gum is used to treat lozenges, emulsions, lotions, sprays, pastes, and throat infections (Tandon and Sharma, 2009). Gum is used to treat diarrhoea, constipation, dysentery, and improved stamina. When treating a throat infection, it serves as a substitute for tragacanth. According to Khan and Khanum (2005), the stem bark of *S. urens* is used to treat spermatorrhoea, leucorrhoea, gastric ulcers, rheumatism, and labour pains. Its pharmacological qualities, including anti-inflammatory, anticancer, antioxidant, antiplasmodic, antitoxic, antidepressant, and antidiabetic actions, have been identified for the different components (Islam et al., 2015).

The incisions of the bark induce gum secretion and find vast applications. Scientific investigations on gum collection have been given by Kuruwanshi et al. (2017), while Dhiman et al. (2019) reviewed the scientific utilization of gum karaya. The present revision focuses on the reported phytochemicals from the species of *Sterculia* and its gum properties besides its applications in various industries.

5.2 PHYSICOCHEMICAL PROPERTIES OF GUM KARAYA

Properties such as solubility, pH, and total proteins of the gum karaya have been studied extensively by Padil et al. (2017). They reported that the gum possesses a mild acidic nature since its pH is 4.8, and this is supported by the reports of Taher (1998). The molecular weight of the gum karaya has been estimated as 9,500,000 D units, determined by Svedberg ultra-centrifugation method by Sao (2012). It may be inferred that the solubility of karaya gum varies with temperature since it dissolves in cold water 10.66% and in hot water 31.34%. In cold water, karaya gum dissolves

DOI: 10.1201/9781003438960-5

the least easily. Instead of dissolving, the gum particle absorbs water and swells to a volume greater than 60 times its initial size, creating a viscous colloidal solution. Acetyl groups in gum karaya's structure are what produce its swelling behaviour (Le Cerf et al., 1990). Given the expectation that solubility will rise with temperature, the gum dissolves more readily in hot water than it does in cold water. However, the samples were only weakly soluble in ethanol and not at all in acetone.

According to Sarah et al. (1998), these suggest that the gum under study has an ionic nature, in contrast to its covalent cousin, which is soluble in solvents that are organic in nature. Due to its ability to ionize and create hydroxyl ions (OH-) and having a greater dielectric constant than acetone, the gum was found to be sparingly soluble in ethanol. As a result, ethanol exhibits some polarity in comparison to acetone, which has a low dielectric constant value. Additionally, the solubility was more in line with the outcome reported by Yadav et al. (2015). It was discovered that 1% of karaya gum was protein. According to Hussein et al. (2006), the majority of the gum was classified as an arabinogalactan due to its extremely low protein concentration.

5.2.1 Phytochemistry of Gum

Kubal and Gralén (1948), Silva et al. (2003), Kohajdová and Karovičová (2009), and others have described karaya gum as a compound of partially acetylated polysaccharides developed as a calcium and magnesium salt. Apart from that, the gum's macromolecules are highly branched cylindrical structures called glicanorhamno-galacto-uritan, with a high molecular mass of about 16×106 Da. The main polysaccharide chain in these structures is made up of D-galacturonic acid units and L-rhamnopyranose groups with α-1,4 and α-1,2 linkages (Rao and Sharma, 1957; Eastwood et al., 1983; Kohajdová and Karovičová, 2009). The main chain's rhamnose residues are 1,4-linked to β-D-galactose units, whereas side chains have 1,2-linkage of β-D galactose or by 1,3-linkage of β-D-glucuronic acid to the galacturonic acid of the main chain (Davidson, 1980; Weiping and Branwell, 2000; Kohajdová and Karovičová, 2009; Mikuš et al., 2014).

Phlobatannin, which is the primary component of the resin, is made up of three phenolic OH groups. Phlobatannin stabilizes beverages without affecting their nutritional value and is useful in removing albuminoids, which are unstable nitrogenous molecules, from them (Herman, 1925). The gum's qualitative phytochemical examination revealed the presence of glycosides, sugars, tannins and phenols, and saponins (Ashwini and Suresh, 2020).

A few secondary metabolites, including β-sitosterol and n-hexacosanol, cycloartenone, cycloartenol, lupeol, and betulin, were identified by Anjaneyulu and Raju (1987) from the stem bark of *S. urens*. The structurally branched gum karaya (*S. urens* – GK) is primarily composed of a partly acetylated carbohydrate polymer (Figure 5.1) bound as a high-molecular-mass (1.6×106 Da) salt of calcium and magnesium (Anderson et al., 1983, 1985; Phillips and Williams, 2001).

The primary structure consists of α-D-galacturonic acid and α-L-rhamnose moieties, which are linked to side chains via 1,2-linkage of β-D-galactose or 1,3-linkage of β -D-glucuronic acid. The majority of rhamnose residues are linked to the core

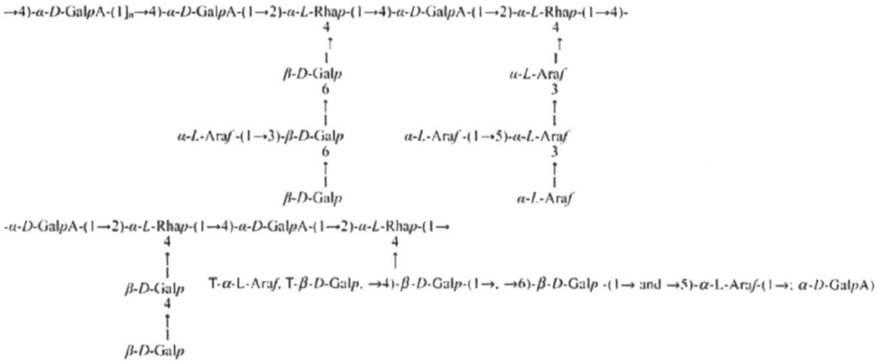

FIGURE 5.1 Molecular structure of polymerizing unit of Sterculia APS from *S. lychnophora* seeds. Source: Wu et al. (2013).

TABLE 5.1
Amino acid composition of three *Sterculia* species

Sl. No	Name of the *Sterculia* species	Amino acids
1	*S. setigera*	Leucine, glutamic acid, and proline
2	*S. villosa*	Valine, proline, glutamic acid, glycine, and threonine
3	*S. urens*	Valine, leucine, proline, and serine

Source: Anderson (1985).

structure through 1,4-linkages of β-D-galactose (Anderson et al., 1983; Brito et al., 2004; de Brito et al., 2005; Le Cerf et al., 1990; Verbeken et al., 2003; Weiping and Branwell, 2000; Whistler and BeMiller, 1993). As shown in Table 5.1, the amino acid content of the gum of four species has been explained by Anderson et al. (1985), with aspartic acid accounting for the majority of the composition.

Galactose (13–26%) and rhamnose (15–30%) are the two main sugar residues identified in GK; the latter is more than other exudate gums. Despite the relatively low protein content (~1%), uronic acid (β-D-glucuronic and D-galacturonic acid) constitutes about 40% of the composition. The reason behind the swelling features of native GK is its high acetyl content (~8%) in its structure, which makes it insoluble in water (Padil et al., 2018). Le Cerf et al. (1990) split the three components of GK and realized that 90% of GK could be dissolved in an aqueous medium following de-acetylation by treatment with NH_3 or sodium hydroxide; in contrast, the native gum (~10%) was soluble in cold water and increased to 30% in hot water (Le Cerf et al., 1990; Stephen et al., 2006).

5.2.2 Structural Elucidation of Gum Karaya

In order to determine the chemical structure of a water-soluble pectic polysaccharide (*Sterculia* APS) that was extracted from boat-fruited *Sterculia* seeds (*S. lychnophora*), 1D/2D nuclear magnetic resonance (NMR) spectroscopy, methylation analysis, and partial acid hydrolysis were used. According to the findings of partial acid hydrolysis, *Sterculia* APS had rhamnose units and galacturonic acids in its backbone, as well as arabinose, galactose, and xylose residues in its branched chains (Wu et al., 2013), besides elucidation of its structural components and side chains, as follows (Figure 5.2):

1. The backbone of *Sterculia* APS was composed of the homogalacturonan (HG) parts referred to as 'smooth' regions consisting of linear linkage of 1,4-α-D-Gal*p*A, and the rhamnogalacturonan-I (RG-I) zones called 'hairy' regions consisting of a→4)-α-D-Gal*p*A-(1 → 2)-α-L-l-Rha*p*-α-(1→backbone partially substituted at *O*-4 of the rhamnose units. The Gal*p*A residues of the backbone were partially methyl esterified and partially *O*-acetylated on C-2 and/or C-3.
2. Side chains of APS mainly consisted of araban and arabinogalactan by the linkages of T-, 1,3-, 1,5-l-Ara*f* and T-, 1,4-, 1,6-, 1,3,6-, 1,3,4-D-Galp, attached to *O*-4 of the backbone rhamnose units. High content of non-reducing terminals of L-Ara*f* and D-Gal*p* indicated that many short side chains were directly attached to the backbone.

5.2.3 Phytochemistry of Leaf Components

The leaves of *S. urens* are reported for the flavonol glycosides, sucrose, β-sitosterol β-amyrin, an ester of terephthalic acid, ascorbic acid, phytol, tannins, stercurens, quercetin, galactose, xylose, rhamnose galacturonic, glucuronic acids, acetyl derivatives, kaemperol derivatives, stercurensin, C-methylchalcone, 3,7,11, 15-tetramethyl-2-hexadecen-1-ol. Gum and mucilage contain aldobiuronic and aldotriouronic acids (Khatoon et al., 1989; Kumbhare and Bhargava, 1999; Brito et al., 2004; Khan

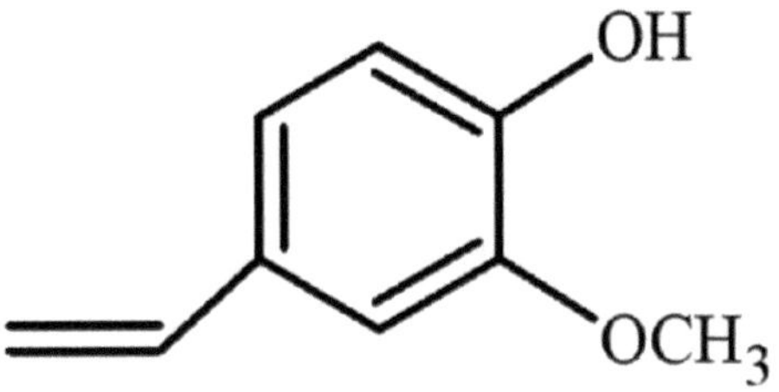

2- methoxy-4-vinylphenol

FIGURE 5.2 2-methoxy-4-vinylphenol.

and Khanum, 2005). Ethanol extract of the leaves of *S. foetida* is reported for the presence of five compounds, such as apigenin-6, puerarin, 5,7,8-tetrahydroxy-4'-methoxyflavone-8-O-β-D-glucoside, 8-di-C-β-D-glucoside, 5,7,8-tetrahydroxy-3',4'-dimethoxyflavone, 5,7,8-tetrahydroxy-4'-methoxyflavone (Mujumdar et al., 2000). β-Sitosterol, β-sitosterol-3-O-β-D-glucoside, and lanosterol were isolated from *S. foetida*, *S. urens*, *S. lychnophora*, and *S. setigera*, (Anjaneyulu and Murty, 1981; Khatoon et al., 1989; Shrivastava and Saxena, 1983; Babalola and Adelakun, 2013; Wang et al., 2003).

Wang et al. (2003) studied and reported alkaloids named sterculinine I and sterculinine II from the ethanol extract of the leaves of *S. foetida*. In addition, 13 known compounds, soya-cerebroside II, 1-O-β-D-glucopyranosyl-(2S,3R,4E,8Z)-2-[(2-hydroxy-icosanoyl) amido]-4, 8-octadecadiene-1,3-diol, kaempferol-3-O-β-D-glucoside, uracil, succinic acid, 2, 4-dihydroxybenzoic acid, daucosterol, β-sitosterol, sucrose, isorhamnetin-3-O-β-D-rutinoside, kaempferol-3-O-β-D-rutinoside, n-butyl-α-D-mannopyranside, and adenosine, were reported from the leaf extract.

The chemical constituents from the leaves of *S. foetida* contained eight compounds, namely, 5,7,8-tetrahydroxy-4'-methoxyflavone-8-O-β-D-glucoside, 5,7,8-tetrahydroxy-4'-methoxyflavone-7-O-β-D-glucoside, quercetin-3-O-β-D-glucoside, apigenin-6, 8-di-C-β-D-glucoside, puerarin, 5,7,8,3'-tetrahydroxy-4'-methoxyflavone, 5,7,8-tetrahydroxy-3',4'-dimethoxyflavone, 5,7,8-tetrahydroxy-4'-methoxyflavone (Xia et al., 2009a). In another study, two new flavonoid glycosides, hypolaetin 4'-methyl ether 8-O- β-D-glucuronide 2''-sulfate and hypolaetin 4'-methyl ether 3'-O- β-D-glucoside, and a new phenylpropanoid glucose ester and 1,6-diferuloyl glucose were characterized from the leaves of *S. foetida* (Xia et al., 2009b).

Phytochemical screening of ethanol extracts (leaves) revealed the existence of tannins, flavonoids, alkaloids, steroids-terpenoids, and glycosides, and phenols, saponins, and anthraquinones were absent (Prastiwi et al., 2018). Pham et al. (2021) reported a new oleanane-type triterpenoid, named stercufoetin A, along with vergatic acid, β-amyrin, oleanolic acid, and maslinic acid, from *S. foetida* using chromatographic methods. Al-Ramamneh et al. (2022) studied the volatile oil components using GC-MS equipment in the ethanolic extracts of the leaves of the chemo-types from Jordanian in *S. diversifolia*, which revealed the presence of major phytochemicals, such as (3β)-Lup-20(29)-en-3-ol acetate (30.97%) and 1-octadecyne (24.88%). Other compounds are squalene (7.19%), germanicol (6.23%), dl-α-tocopherol (5.24%), heptacosane (4.41%), phytol (3.54%), and pentacosane (2.89%).

The major phytoconstituents reported in methanolic extract of *S. villosa* (stem bark) included terpenoids, phytosterols, esters, acids, and other organic compounds, such as mebutamate, octadecanoic acid, 2-hydroxy-1,3-propanediyl, estradiol, methotrexate, daucol, meprobamate, hexadecanal, glycerol-1-palmitate, cyclohexane, and eicosyl (Lyzu et al., 2022).

5.2.3.1 Stem Bark

Calamenene-type sesquiterpenoids purified from the methylene chloride fraction of *S. tavia* bark, named tavinin A, epi-tavinin A, and mansonone G, demonstrated antiproliferative activity on A2780 ovarian cancer cell line (Dai et al., 2012; Thabet et al., 2018). Epicatechin, procyanidin B2, and catechin were identified from the stem and root bark of *S. tragacantha* (Orisakeye and Olugbade, 2014).

The stem bark of *S. foetida* was reported for eight compounds consisting of cleomiscosin A, cleomiscosin C (coumarinolignans), quercetin, kaempferol 3-*O*-β-D-xylopyranoside (flavonols), lupeol, lupenone, betulinic acid, and taraxerol (triterpenes) of Binh Thuan province, Vietnam (Pham et al., 2022).

5.2.4 Phytochemicals of Roots

Roots contain coumarin and scopoletin (Khatoon et al., 1989; Kumbhare and Bhargava, 1999; Brito et al., 2004; Khan and Khanum, 2005). Shukla et al. (2020) reported the presence of alkaloids, steroids, glycosides, flavonoids, phenolics, tannins, and saponins using qualitative analysis.

5.2.5 Seed Components

A straight-chain ester, nonacosanoic acid, methyl ester, was isolated for the first time from acetone extract of *S. guttata* (Katade et al., 2008). The chemical composition of kernel flour of *S. striata* reported about 25.8% lipid content. Physicochemical parameters such as acid, iodine, peroxide, and saponification values were 0.82 (% as oleic acid), 69.2 (g iodine/100 g oil), 4.20 (m eq./kg), and 136.1 (mg. KOH/g oil), respectively. With respect to fatty acid composition, the oil contained 36.2, 43.7, and 10.9% saturated, monounsaturated, and polyunsaturated fatty acids, respectively. Palmitic acid (31.9%), oleic acid (41.7%), and linoleic acid (10.73%) were the principal saturated, monounsaturated, and polyunsaturated fatty acids. Two cyclopropenoid fatty acids, namely, sterculic and malvalic acid, were identified at a concentration of 5.3 and 2.3%, respectively (Zeomar et al., 2008).

The fatty acid composition in fixed oil extracted from *S. foetida* seed constituted seven fatty acids, such as methyl esters of tetradecanoic acid (myristic acid) (1.65%), hexadecanoic acid (palmitic acid) (11.87%), 9-octadecenoic acid (oleic acid) (20.50%), 7, 10-octadecadienoic acid (linoleic acid) (20.50%), and 8-(2-octacyclopropen-1-yl)octanoic acid (sterculic acid) (6.76%), characterized using GC-MS analysis by Kale et al. (2009).

Roasted and natural seed oils were found to include malvalic acid, sterculic acid, and cyclopropenoic fatty acids. As has been noted for *S. mexicana* and *S. foetida*, sterculic acid constituted the majority of the seed oil. Steric acid (19:1 cyclo) was the main constituent of the seed oils from *S. apetala* and *S. mexicana*; the former had the largest amount (56.3 ± 0.91 vs. 51.3 ± 0.29 in *S. mexicana*, $p < 0.02$; 9% greater). Malvalic acid (18:1cyclo), the other cyclopropenoid fatty acid (CPFA), was found in trace amounts in *S. mexicana* and *S. apetala* (1.1 ± 0.05 and 1.3 ± 0.06, respectively). In terms of other fatty acids, oleic acid (18:1), stearic acid (18:0), and palmitic acid (16:0) were the most prevalent. There were small but noteworthy variations in oleic acid. Oleic acid (C18:1) (8.8 ± 0.09 in *S. mexicana* vs. 9.5 ± 0.16 in *S. apetala*, $p < 0.04$) and linoleic acid (C18:2) (3.8 ± 0.06 in *S. mexicana* vs. 2.9 ± 0.11 in *S. apetala*, $p < 0.01$) showed slight but statistically significant differences. The percentages of stearic acid (18:0) and palmitic acid (C16:0) were found to be comparable in both species.

S. apetala and *S. mexicana* had higher sterculic acid contents than other *Sterculia* species (56.3% and 51.3% vs. 11.3% in *S. tomentosa*, 30.2% in *S. tragacantha*, 5.3% in *S. striata*, 4% in *S. alata*, 5.8% in *S. guttata*, and 3.2% in *S. villosa*), but lower malvalic acid contents were observed (1.3% and 1.1% vs. 5.8% in *S. tomentosa*, 5.1% in *S. tragacantha*, 2.3% in *S. striata*, 17.6% in *S. alata*, 2.1% in *S. guttata*, and 2.5% in *S. villosa*) (Miralles et al., 1993; Nitao et al., 2008; Badami et al., 1980). On the other hand, the profiles of fatty acids resembled those of *S. foetida* seed oil more (Chaves et al., 2004; Herrera-Meza et al., 2014).

Chemical and functional characterizations of the meal of gum karaya (*S. urens*) whole and dehulled-defatted seed were investigated. Dehulled-defatted seed meal (DDSM) contained 40.7% of protein as compared to 20.4% in the whole seed meal (WSM). Essential minerals of Ca 39.5, Fe 4.4, K 9.6, and P 995 mg/100 g were reported in DDSM (Galla and Dubasi, 2010).

It has been reported that processed seeds meal enhances foam-forming capability and stability. Since emulsification depends on the ability to hold oil and water, proteins found in seeds provide good emulsifying agents (Elizalde et al., 1991; Galla and Dubasi, 2010). At pH 12, protein solubility reaches its maximum of 93.97%, and as pH drops, it drops to 26.41%. Protein solubility at the isoelectric point (pH 6) is further increased by the addition of salts (NaCl) (Galla and Dubasi, 2010). When placed in an alkaline media, processed seed meal has a lower buffer capacity at acidic pH values. Due to its lower initial moisture content (5.16%), which abruptly climbs to 70% when exposed to greater humid circumstances, dehulled and defatted seed meal has been discovered to be non-hygroscopic in nature. Cotyledonary flour, a rich source of protein and fat, when combined with other seed meals, can be used to make nutritious supplements, according to FAO/WHO. *S. urens* is a viable source for the biodiesel and soap industries due to its high lipid content and good quantities of saturated fatty acids (Galla and Dubasi, 2010).

The kernel, which constitutes up to approximately 56% of the seed, is the edible portion of the seed. In addition to being used to make soap, seed oil is edible (Yesodharan and Sujana, 2007). Previous analyses have examined both functional and chemical features (Galla and Dubasi, 2010). There have been reports of a high protein (30%), carbohydrate (28%), and fat and lipid (26%) content (Galla and Dubasi, 2010). The protein in seeds is a mixture made up of both essential and non-essential amino acids. The nutritional value of the seeds is strengthened by the presence of the important amino acids leucine, isoleucine, and threonine (Galla and Dubasi, 2010).

The amino acid and fatty acid compositions of *S. urens* seeds are reported. The cotyledons were found to be rich in protein (30.88%) and lipids (39.2%). The major amino acids in defatted *S. urens* cotyledon flour (DSCF) were determined as glutamic acid, arginine, and aspartic acid. Cysteine, methionine, tyrosine, and histidine were observed in minor quantities. The ratio of essential to non-essential amino acids was observed to be 0.45. Among the essential amino acids, isoleucine was found to be higher than the reported FAO/WHO requirements. The GC-FID and GC-MS analysis revealed that the major fatty acids of the total lipid were stearic acid (31.72%), linoleic acid (28.83%), and palmitic acid (26.79%). Eicosadienoic acid (4.98%) and eicosatrienoic acid (2.96%) were also found in the total lipid (Galla et al., 2012).

Two cerebrosides, named 1-O-β-D-glucopyranosyl-(2S,3R,4E,8Z)-2-[(2-hydroxyoctadecanoyl) amido]-4,8-octadecadiene-1,3-diol (demonstrated for neuroprotective activity on SH-SY5Y cells) and soya-cerebroside I, were isolated from the seeds of *S. lychnophora*. (Wang et al., 2013). Sterculic acid (8-(2-octyl-cyclopropen-1-yl)-octanoic acid), a thermal unstable component, was detected in the kernel oil from *S. foetida*, and an unusual fatty acid, such as sterculynic acid, 7-[2-(8-nonynyl)-cyclopropen-1-yl]-heptanoic acid), from the seed oil of *S. alata* (Avato and Tava, 2022).

5.2.6 Aerial Parts Phytochemical Composition

Lowry (1971) reported anthocyanin, pelargonidin-3-arabinoside from the follicles of *S. parviflora*. In addition to the customary uses of its various plant components, attempts have been made to substantiate these uses through scientific means. From the gas chromatography–mass spectrometry (GC–MS) analysis of *Sterculia* leaves, Nanadagopalan et al. (2015) confirmed the identity of 27 compounds. The major ones are 2-methoxy-4-vinylphenol (Figure 5.2; peak area: 5.27%), 5(2H)-oxazolone, 4-(phenylmethyl) (peak area: 6.97%), 2, 3-dihydro-3, 5-dihydroxy-6-methyl-4H-pyran-4-one (peak area: 6.78%), megastigmatrienone (peak area: 5.93%), and 2-methoxy-4-vinylphenol (Figure 5.2; peak area: 5.27%). Organic substances like phytol are frequently employed as aroma ingredients for both cosmetic and non-cosmetic products, such as detergents, cleansers, and fragrances. Sucrose's presence suggests that it has sweetening qualities that increase nutritional value. The main tasting ingredient in pineapple and miracle berries is 2,4-dihydroxy-2, 5-dimethyl-3(2 H)-furan-3-one, an acetyl formoin intermediate (Van Cott et al., 1954). It serves as the primary intermediary in the synthesis of 4-hydroxy-2, 5-dimethyl-3(2 H)-furanone, the main flavouring that enhances the sensory qualities of food products that have undergone heat processing. 2, 5-dimethyl-2, 4-dihydroxy-3(2 H)-furanone is a flavouring agent that is utilized in the food and pharmaceutical industries and has a significant impact on gustation (Chukwu, 2017). Strong antioxidant properties are exhibited by 2,3-dihydro-3, 5-dihydroxy-6-methyl-4H-Pyran-4-one (DDMP) (Ban et al., 2007).

Research indicates that the soyasaponin derivative's DDMP moiety is primarily responsible for scavenging reactive oxygen radicals (Yoshiki and Okubo, 1995). With their five-membered heterocyclic nucleus, 5(2 H)-oxazolone and 4-(phenylmethyl) are important building blocks for the synthesis of biologically active molecules, such as amino acids, α-keto acids, aryl acetic acids, and peptides. It is known that oxazolones have potential uses in pharmacology, photochemistry, and photophysical applications. According to Mesaik et al. (2004) and Bala et al. (2011), their intermediates have anti-HIV, anticancer, anti-tumour, analgesic, anticonvulsant, pesticidal, antimicrobial, antiangiogenic, and cardiotonic properties. Burley tobacco's primary flavouring component, megastigmatrienone, is known for its sweet scent and is fictitiously odoriferous (Ohloff, 1978).

The addition of tobacco aroma to wine indicates the presence of megastigmatrienone (Slaghenaufi and Ugliano, 2018). Wine's concentration of megastigmatrienone increases as it ages (Slaghenaufi et al., 2014). 2-methoxy-4-vinylphenol (2M4VP) is a naturally occurring phenolic substance that is employed as a flavouring. There have been reports of 2M4VP in curry, apple, peanut, clove, and buckwheat. It has anticarcinogenic properties and controls the expression of cyclin-CDKs in the cell cycle. According to research, 2-methoxy-4-vinylphenol (Figure 5.2) has strong anti-inflammatory effects because it helps regulate NF-κB, MAPK, and histone acetylation (Jeong et al., 2011).

5.2.7 Seed Essential Oils

Amuthavalli et al. (2021) reported 13 phytochemicals from the seed powder of *S. foetida* methanolic extract. The pharmacologically active major compounds present in the seeds were flavanthrone and flavone (anticancer activity), 5-undecyne, methyl abietate (antioxidant), 2-cyclohexen-3-ol-1-one,2(11-phenylundecanoyl) (antidiuretic), androstan-6–01-17-one,3-acetoxy-5Achloro (vasodilator), and testosterone cypionate (anti-inflammatory), besides oleic acid, 10-octadecenoic acid methyl ester, flavanthrone, 5-undecyne, palmitic acid, retinal,9-cis, methyl abietate, testosterone cypionate, etc.

The chicha (*S. striata*) seed oil is rich in fatty acids containing a cyclopropene ring (41%) in their chain, mainly sterculic (octadec-9,10-methylene-9-ene) and malvalic (heptadec-8,9-methylene-8–1-ol) acids (Chaves et al., 2004; Mangas et al., 2012). Species like *S. foetida* (Vickery, 1980) and *S. tragacantha* (Pawlowsky et al., 1972) show also a high level of cyclopropenoid fatty acids (CPFA).

5.3 CHEMICAL NATURE OF THE GUM KARAYA AND ALLIED SPECIES

Commercial amounts of tapped exudate gum are typically harvested from the *S. urens* tree, which is primarily found in central and northern India. *S. villosa* in India and Pakistan and *S. setigera* in Mali, Senegal, and the Sudan (Wang, 2000) are additional sources of gum (Wareing, 1997). There is minimal difference in gum karaya's chemical makeup throughout the tree species that are exploited. Working with *S. urens*, *S. villosa*, and *S. setigera*, Anderson et al. (1983) discovered few significant chemical differences. Gum karaya has a molecular weight of up to 16 million Da and is a complex, branching, partly acetylated polysaccharide (Le Cerf et al., 1990). In different amounts, hydrolysis produces d-glucuronic acid, d-galacturonic acid, d-galactose, and l-rhamnose (Meer, 1980). Compared to other exudate gums, gum karaya contains a significantly higher l-rhamnose concentration, which may be important for identification. The polysaccharide has between 8 and 14% acetyl groups and about 40% uronic acid residues. Although acetyl groups facilitate swelling, they also inhibit the gum from completely dissolving in water (Wareing, 1997). Gum that has been treated with alkali will chemically deacetylate, changing its properties and increasing its solubility in water. Gum karaya's structure consists of side chains of glucuronic acid and a core chain of galactose, rhamnose, and galacturonic residues. The exudate is present as a salt mixture of calcium and magnesium. Although the amino acid makeup of different *Sterculia* species varies, around 1% of proteinaceous material has been found (Mayes, 2010).

5.3.1 Solubility, Viscosity of Gum-Karaya

Gum karaya is the least-soluble of the commercial exudates due to the presence of acetyl groups, yet at low concentrations, it may quickly absorb cold water and produce a thick colloidal dispersion. The gum makes thick, smooth, gel-like pastes that are perfect for suspending food at high concentrations of 3–5%. These pastes are spreadable. Particle size affects the rate of hydration; finer mesh material hydrates more quickly than coarse mesh material, giving the appearance of homogenous dispersions. When coarse-powdered gum is added to grainy dispersions, high shear produces smooth solutions with a lower viscosity. Gum karaya mucilages become more soluble when heated because of conformational changes in the polymers, but viscosity is permanently lost. Gum can be hydrated in cold water to a maximum of 4%. At concentrations as high as 20%, heating under pressure generates smooth, transparent, colloidal solutions (Glicksman, 1983). Raising the pH can help pre-hydrated gum dispersions become more soluble. After that, the polymer exhibits random coil behaviour, resulting in greater viscosities with a cohesive and ropy texture (Le Cerf et al., 1990). According to Anderson et al. (1983), *S. setigera* gum's physicochemical characteristics are more or less the same as those of the other *Sterculia* species. Gum karaya from African sources (*S. setigera*) tends to offer more ropy textures since it has a somewhat-lower acetyl content than gum karaya from Indian sources (*S. urens*). According to the grade, gum karaya dispersions have viscosities ranging from 120 to 400 mPa·s for 0.5% dispersions to over 10,000 mPa·s for 3% dispersions. As one ages, viscosity in powdered form reduces. The viscosity of finely ground powder decreases more quickly after grinding than that of coarse powder or the whole exudate. Viscosity loss can be caused by either high storage temperature or high storage humidity. Viscosity is also decreased when gum solutions are boiled for longer than two minutes (Wang, 2000).

5.3.2 Rheological Properties of Gum Karaya

The gum solutions are thixotropic, according to rheological research, and as a consequence, the apparent viscosity is dependent on the solution history (Mayes, 2010). Sahu et al. (2019) researched the significant rheological and physicochemical characteristics of karaya gum. Following the established procedures for other comparable food materials, attempts have been made to determine the moisture content, bulk density, true density, tap density, bulkiness, porosity, Hausner's ratio, Carr's compressibility index (%), angle of repose (°), and coefficient of friction (n) on glass, mild steel, plywood, and rubber surfaces. For the study, lumps of karaya gum taken from the Chhattisgarh forest were used. Depending on what the technique required, both grits and powder samples were utilized to determine the qualities. The following physical and chemical properties were measured: bulk density: 0.79 ± 0.02 (g/cm^3); true density: 1.50 ± 0.17 (g/cm^3); bulkiness: 1.26 ± 0.03 (cm^3/g); porosity: 46.76 ± 6.12 (%); Hausner ratio: 1.14 ± 0.06; Carr's compressibility index: 12.31 ± 4.19; angle of repose: 47.27 ± 2.91°; coefficient of friction: 0.24 ± 0.03 N (glass); 0.43 ± 0.04 N (mild steel); 0.28 ± 0.01 N (plywood); and 0.47 ± 0.04 N (rubber), respectively. The values of nitrogen (%), protein (%), water activity, pH, refractive index, ash content, and water-holding capacity (per 100 ml) were found to be 4.62, 4.26, 1.336, 0.651, 84.76, 0.16, and 1.06, respectively. Karaya gum is insoluble in acetone, chloroform, and ethanol but soluble in both hot and cold water. At room temperature of 24–26°C, it was discovered that the viscosity of 1% gum solution ranged from 619 to 1,286 cp in the spindle rotating speed range of 20 to 100 revolutions per minute (Sahu et al., 2019).

5.3.3 Heat Stability

In heated gum karaya dispersions, irreversible polymer conformation changes take place, leading to a permanent loss of viscosity. Nonetheless, concentrations of up to 20% can be created under carefully regulated pressure and temperature conditions, resulting in extremely viscous solutions (Mayes, 2010).

5.3.4 Water Binding and Adhesion Properties

Gum karaya has the ability to expand and absorb water up to 60 times its initial volume. Gum karaya produces pastes with good adhesion qualities at concentrations of 20–50%. The main applications for these pastes were in colostomy bag sealing rings, dental adhesives, and bulk laxatives. Nonetheless, current dental research indicates that natural teeth suffer when they are in constant contact with acidic gum tissue (Mayes, 2010).

5.3.5 pH Stability

Because of the acid residues, acidic by nature, gum karaya has a pH that ranges from 4.4 to 5.2. De-acetylation results from increasing pH, and the solution's texture becomes cohesive and ropy. If alkali is introduced to the water before the gum is added, this phenomenon does not happen. Additionally, the presence of electrolytes retards the maximum viscosity. It is therefore usually preferable to add the gum to the water before adding the salts and buffers in food applications where a rapid development of maximal viscosity is needed. Because gum karaya is acid-stable, it works best for thickening dressings and sauces with acidic bases that are made with little to no heat processing (Mayes, 2010).

5.4 APPLICATION OF CHEMICAL TECHNOLOGY FOR PREPARATION OF BENEFICIAL PRODUCTS

Multiple teams of researchers have extensively examined the physicochemical, related to structure, and rheological characteristics of GK (*S. urens*), as well as its occurrence and production in food and non-food applications (Vinod et al., 2010, 2015). GK is a high-molecular-mass polysaccharide with a branching structure and a partial acetylation of ~16 × 106 Da. It falls within the category of pectic tree gums, or 'substituted rhamnogalacturonoglycans'. About 60% of the sugars in this gum are neutral (rhamnose and galactose), 40% are acidic (glucuronic acid and galacturonic acids), and 8% are acetyl groups (Vinod et al., 2010, 2015). With respect to its high viscosity, acid stability, and capacity to bind water and create suspensions, GK is a good emulsification agent. Natural hydrocolloid-based polymeric surfactants are becoming more and more important because of their superior physicochemical qualities, availability, biodegradability, and lack of toxicity. Octenylsuccinic anhydride (OSA) and dodecenylsuccinic anhydride (DDSA), two long-chain alkyl and alkenyl dicarboxylic acid anhydrides, have recently been employed to synthesize derivatives of starch, inulin, and konjac glucomannan. Furthermore, the stabilization, encapsulation, interfacial, thermal, emulsification, and rheological properties have been better promoted by these OSA and DDSA derivatives of natural polysaccharides.

5.5 APPLIED CHEMICAL TECHNOLOGY FOR IMPROVED PRODUCTS FROM GUM KARAYA

5.5.1 Preparation of De-Acylated Gum Karaya (DGK)

To obtain fine and consistent samples, the gum of *S. urens* was ground into a powder using a high-speed mechanical blender and then sieved through a bin with a mesh size of 250 μm. GK powder was precisely weighed in 1 g amounts and poured into spotless glass beakers holding 1 L of deionized water. The gum solutions were put on magnetic stirrers and left to gently swirl overnight at room temperature. To extract any remaining material, they were then left to stand at room temperature for 12 hours. To get clear solutions, the gum solutions that were produced were centrifuged. One volume of 1 M NaOH was mixed with three volumes of each gum solution to deacetylate them. One volume of 1 M HCl was added to the solution to neutralize it and bring its pH down to 7.0 after it had been incubated for 6 hours at room temperature (with light stirring on a magnetic stirrer). To get rid of any remaining salts, the resultant DGK solution was dialyzed repeatedly against deionized water. Dialyzing the centrifuged DGK solution against deionized water was done. Following dialysis, any remaining precipitates in the gum solution were removed by centrifuging it; the resulting clear solutions were then freeze-dried and kept for later use (Vinod et al., 2015).

5.5.2 Preparation and Characterization of DDSA Derivative of DGK

Esterification of DGK. Figure 5.2 shows the plan for esterifying gum karaya (deacetylated gum karaya) using derivatives of decenylsuccinic anhydride gum karaya (DDSA). The concentration ratio of DDSA and DGK, pH, concentration of NaOH solution, reaction duration, and temperature were the main factors influencing the creation of DDSA derivatives of the DGK. According to Vinod et al. (2015), the current investigation maintained the DDSA-DGK (Figure 5.3) ratio at 3:1, the reaction mixture's pH between 8 and 9 using 3% NaOH, and the temperature between 25 and 30°C for eight hours. The creation of gums modified with DDSA has the potential to serve as stabilizers and antibacterial agents for both food and non-food applications (Vinod et al., 2015).

5.5.3 Polyacrylamide Hybridized with Gum Karaya Product (Hydrogel)

The production and application of a nickel sulphide nanoparticle-based hydrogel and gum karaya (GK) grafted with polyacrylamide (PAAm) to efficiently remove

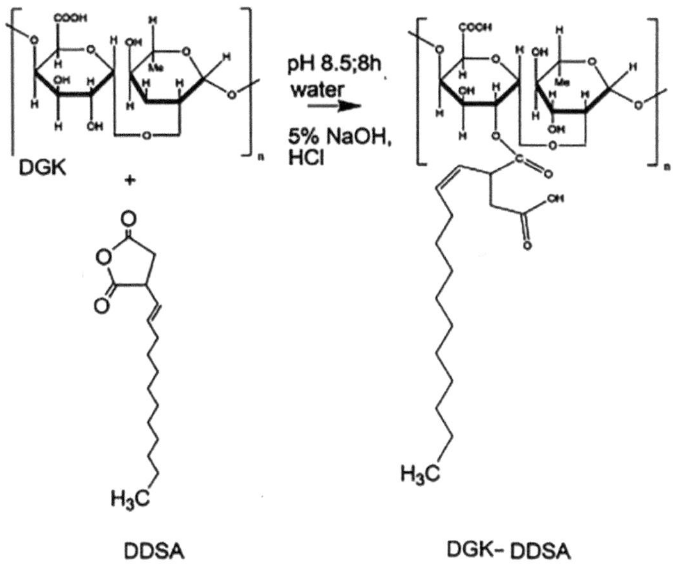

FIGURE 5.3 Representative reaction among DDSA and DGK and formation of the DDSA-DGK product.

Source: Vinod et al. (2015).

rhodamine 6G dye (R6G) from aqueous solution. Initially, the graft copolymerization method was used to create the hydrogel polymer of the GK with the PAAm. The second stage involved the in situ incorporation of nickel sulphide nanoparticles into the hydrogel polymer matrix. The adsorption–desorption cycle was repeated five times with the hydrogel nanocomposite. As a result, according to Kumar et al. (2016), the nanocomposite hydrogels have demonstrated their potential for eliminating cationic dyes from aqueous solutions.

5.5.4 Gum Karaya as an Excellent Binder for Silicon-Based Anodes

Gum Karaya was studied as a binder for high-performance silicon-based anodes. It is made up of polysaccharides and glycoproteins. Gum karaya electrodes exhibit high electrochemical performance due to their many polar function groups, which result in superior binding strength and good mechanical qualities facilitated by their multi-branched structure and glycoproteins. Gum karaya electrodes have a lithiation capacity and can sustain a reversible capacity of 1,000 mA h g/L for over 1,200 cycles at 4.0 A g/L. Gum karaya electrodes demonstrate good cycle stability over 100 cycles when applied to the Si/graphite composite electrode for practical consideration, whereas other electrodes deteriorate quickly in the first few cycles. Gum karaya electrodes exhibit improved electrochemical performance due to their multi-branched structure, which provides a large number of polar groups and abundant multidimensional bonding sites; uniform SP dispersion forms an effective conductive network; and the protein part and multi-branched structure combine to form a stable electrode structure. Gum karaya is a great option as a binder for Si-based anodes due to its additional benefits, which include its natural abundance, low cost, eco-friendliness, and mature industrial process (Bie et al., 2017).

5.5.5 Solid Dispersions from Gum Karaya

Gum karaya was ground into a powder in a porcelain thimble and heated to 120°C for two hours. The resulting powder was sieved through a mesh (#80) that is utilized for medication-dispersion media. In modified gum karaya, Aceclofenac, a medication used as an anti-inflammatory and antirheumatoid agent, has been blended and distributed in different ratios, and its standardization assessed. Additionally, the dispersion media's drug-releasing ability is shown to be useful for controlled release (Arora et al., 2011). Khan et al. (2010) demonstrated similar experiments with ambroxol hydrochlorides.

5.6 FUTURE PROSPECTIVE WITH SPECIAL REFERENCE TO GUM KARAYA SECONDARY METABOLITES

Since the natural gum industry has become uncertain regarding its prospects for the future, a thorough economic analysis of the country's and the world's trade is required. The industry prefers synthetic products due to the unpredictability of natural gum supply and cost; however, unstable oil prices, decreased production, and high costs of synthetic material create a promising future for natural gums and resins. Despite the competition from synthetic products, natural gum and resins are preferred in certain industries because they are superior; the tapping methods used are harsh and harmful to the plants, frequently resulting in their death; and the outdated technology needs to be updated in order to ensure sustainable yield and quality control.

There is an immediate need for research institutions, universities, and non-governmental organizations to work together to improve the industry's tapping, collection, processing, grading, classification, and marketing, among other elements. When it comes to using natural gums and resins, R&D is partially found in certain countries. The industry is entirely dependent on conventional research as well as some ad hoc work conducted by people. It is important to start investigating genetic improvement and species selection for the production of gums and resins, as this could result in the construction of these species' plantations. The gum and resin industries can keep rural residents from migrating to towns and cities by giving them jobs and a reliable source of extra income.

REFERENCES

Al-Ramamneh, E.A.M., A.M. Ghrair, A.K. Shakya, K.Y. Alsharafa, K. Al-Ismail, S.Y. Al-Qaraleh, J. Mojski and R.R. Naik. 2022. Efficacy of *Sterculia diversifolia* leaf extracts: Volatile compounds, antioxidant and anti-inflammatory activity, and green synthesis of potential antibacterial silver nanoparticles. *Plants* 11(19): 2492. doi: 10.3390/plants11192492.

Amuthavalli, A. and T. Ramesh. 2021. Phytochemical, fluorescence and GC-MS analysis of methanolic extract of *Sterculia foetida* L. seeds. *Intern. J. Environ. Agricult Biotechnol.* 6(1): 46–53. doi: 10.22161/ijeab.61.7

Anderson, D.M.W., J.F. Howlett and C.G.A. McNab. 1985. The amino acid composition of the proteinaceous component of gum karaya (*Sterculia* spp.). *Food Additives Contaminants* 2: 153–157.

Anderson, D.M.W., C.G.A. McNab, C.G. Anderson, P.M. Braown and M.A. Pringuer. 1983. Studies of uronic acid materials, Part 58: Gum exudates from genus *Sterculia* (gum karaya). *Intern. Tree Crops J.* 2: 147–154. doi:10.1080/01435698.1983.9752749

Anjaneyulu, A. and V. Murty. 1981. Two rare tetramethyl ethers of quercetin from *Sterculia foetida* Linn. *Indian J. Chem.* B 20: 87–88.

Anjaneyulu, A. and S. Raju. 1987. Terpenoids and phenolics from the bark and heartwood of *Sterculia urens* Roxb. *J. Indian Chem. Soc.* 64: 323–324.

Arora, G., K. Malik, J. Sharma, M. Nagpal and I. Singh. 2011. Preparation and evaluation of solid dispersions of modified gum karaya and aceclofeac: Controlled release application. *Der Pharmacia Sinica* 2(2): 142–151.

Ashwini, M.K. and B.R. Suresh. 2020. The phytochemical analysis of gum samples from Gadchiroli District, Maharashtra, India. *J. Emerging Technologies and Innovative Res* 7(4): 1536–1541.

Avato, P. and A. Tava. 2022. Rare fatty acids and lipids in plant oilseeds: occurrence and bioactivity. *Phytochem. Rev.* 21, 401–428. https://doi.org/10.1007/s11101-021-09770-4.

Babalola, I.T. and E.A. Adelakun. 2013. Isolation of stigmast-5-en-3b-ol (β-sitosterol) from dichloromethane extract of *Sterculia setigera* leaves (Sterculiaceae). *Arch. Appl. Sci. Res.* 5: 16–19.

Badami, R.C., K.C. Patil, Y.V. Subbarao, G.S.R. Sastry and V.G.K. Rao. 1980. Cyclopropenoid fatty acids of *Sterculia* oils by gas liquid chromatography. *Fette Seifen Anstrichm* 82: 317–318. http://dx.doi.org/10.1002/lipi.19800820806.

Bala, S., M. Saini, S. Kamboj, G. Uppal and A. Kumar. 2011. Biodiversity of oxazolone derivatives in medicinal chemistry: A review. *Asian J. Res. Chem.* 4: 685–694.

Ban, J.O., D.Y. Yuk, K.S. Woo, T.M. Kim, U.S. Lee, H.S. Jeong, D.J. Kim, Y.B. Chung, b.Y. Hwang, K.W. Oh and J.T. Hong. 2007. Inhibition of cell growth and induction of apoptosis via inactivation of NF-κB by a sulfur compound isolated from garlic in human colon cancer cells. *J. Pharmacol. Sci.* 104: 374–383. https://doi.org/10.1254/jphs.FP0070789

Bie, Y., J. Yang, Y. Nuli and J. Wang. 2017. Natural karaya gum as an excellent binder for silicon-based anodes in high-performance lithium-ion batteries. *J. Mater. Chem. A* 5: 1919–1924.

Brito, A.C.F., D.A. Silva, R.C.M. de Paula and J.P.A. Feitosa. 2004. *Sterculia striata* exudate polysaccharide: Characterization, rheological properties and comparison with *Sterculia urens* (Karaya) polysaccharide. *Polymer International* 53(8): 1025–1032.

Chaves, M.H., A.S. Barbosa, J.M.M. Neto, S. Aued-Pimentel and J.H.G. Lago. 2004. Chemical characterization of the oil of *Sterculia striata* St. Hil. et Naud nuts. *Quim Nova* 27(3): 404–8.

Chukwu, C.J. 2017. Isolation, purification and characterization of flavor compounds in the fruit pulp of *Sysepalum dulcificum*. *FUNAI J Sci Technol* 3: 67–84.

Dai, Y., L. Harinantenaina, P.J. Brodie, M.W. Callmander, S. Randrianasolo, E. Rakotobe, V.E. Rasamison and D.G. Kingston. 2012. Isolation and synthesis of two antiproliferative calamenene-type sesquiterpenoids from *Sterculia tavia* from the Madagascar rain forest. *Bioorg Med Chem.* 20(24): 6940–6944. doi: 10.1016/j.bmc.2012.10.012.

Davidson, R.L. 1980. *Handbook of water-soluble gums and resins*. McGraw-Hill, London.

De Brito, A.C.F., M.R. Sierakowski, F. Reicher, J.P.A. Feitosa and R.C.M. De Paula. 2005. Dynamic rheological study of *Sterculia striata* and karaya polysaccharides in aqueous solution. *Food Hydrocoll.* 19: 861–867. https://doi.org/10.1016/j.foodhyd.2004.10.035

Dhiman, M., A. Singh and M.M. Sharma. 2019. A review on *Sterculia urens* Roxb.: A boon to the livelihood for tribal people and industry. *Industrial Crops Prod.* 130: 341–351.

Elizalde, B.E., D. Giaccaglia, A.M.R. Pilosof and G.B. Bartholomai. 1991. Kinetics of liquid drainage from protein–stabilized foams. *J. Food Sci.* 56: 24–26. https://doi.org/10.1111/j.1365-2621.1991.tb07966.x.

Eastwood, M.A., W.G. Brydon and D.M.W. Anderson. 1983. The effects of dietary gum karaya (*Sterculia*) in man. *Toxicol. Lett.* 17: 159–166. https://doi.org/10.1016/0378-4274(83)90052-8.

Galla, N.R. and G.R. Dubasi. 2010. Chemical and functional characterization of Gum karaya (*Sterculia urens* L.) seed meal. *Food Hydrocolloids* 24(5): 479–485. https://doi.org/10.1016/j.foodhyd.2009.12.003

Galla, N.R., P.P. Rao and A. Satyanarayana. 2012. Chemical, amino acid and fatty acid composition of *Sterculia urens* L. seed. *Food Hydrocolloids* 28: 320–324.

Glicksman, M. 1983. Gum karaya (*Sterculia* Gum). In: M. Glicksman (ed), *Food hydrocolloids*. Vol. II. CPC Press, New York, pp. 39–47.

Herman, H. 1925. Manufacture of Stable Beverages. US Patent 1:550,359.

Herrera-Meza, S., A.J. Martínez, M.G.Sánchez-Otero, M.R. Mendoza-López, O. García-Barradas, G.R. Ortiz-Viveros, 2014. Fatty acid composition and some physicochemical characteristics of *Sterculia apetala* seed oils. *Grasas y Aceites* 65(3): e039. doi:10.3989/gya.0223141.

Hussein, M.S. 2006. Comparative study on some Sudanese gums: Hashab (*Acacia senegal*), talha (*Acacia seyal*) and Karaya (*Sterculia urens*). M.Sc. Thesis. Faculty of Science, University of Kordofon.

Islam, M.T., M.V.O.B. de Alencar, K. da Conceição Machado, K. da Conceiçao Machado, A.A. de Carvalho Melo-Cavalcante, D.P. de Sousa and R.M. de Freitas. 2015. Phytol in a pharma-medico-stance. *Chem-Biol. Interact.* 240: 60–73. https://doi.org/10.1016/j.cbi.2015.07.010

Jeong, J.B., S.C. Hong, H.J. Jeong and J.S. Koo. 2011. Anti-inflammatory effect of 2-methoxy-4-vinylphenol via the suppression of NF-κB and MAPK activation, and acetylation of histone H3. *Arch. Pharmacol. Res.* 34: 2109–2116. https://doi.org/10.1007/s12272-011-1214-9

Kale, S.S., V. Darade and H.A. Thakur. 2009. Analysis of fixed oil from *Sterculia foetida* Linn. *Int. J. Pharm. Sci. Res.* 2(11): 2908–2914.

Katade, S., M. Deshmukh, U. Phalgune, S. Biswas and N. Deshpande. 2008. Isolation of straight chain alcohol and ester from *Sterculia guttata*. *Asian J Chem* 20(1): 308–312.

Khan, I.A. and A. Khanum. 2005. *Medicinal and aromatic plants of India*. Ukaaz Publications, Hyderabad.

Khan, K.A.A., Md. Sarfaraz, B.R. Vinod and S.R. Reddy. 2010. Formulation and *in vitro* evaluation of Karaya gum based controlled release matrix tablets of ambroxol hydrochloride. 2010. Poster Presentation in Conference: 62nd Indian Pharmaceutical Congress, held at Manipal University.

Khatoon, F., A.M. Khabiruddin and W.H. Ansari. 1989. Flavonol glycosides from the leaves of *Sterculia urens* Roxb. *J. Indian Chem. Soc.* 66: 287–288.

Kohajdová, Z. and J. Karovičová. 2009. Application of hydrocolloids as baking improvers. Chem. Pap. 63: 26–38. https://doi.org/10.2478/s11696-008-0085-0.

Kubal, J. and N. Gralén. 1948. Physicochemical properties of karaya gum and locust bean mucilage. *J. Colloid Sci.* 3: 457–471. https://doi.org/10.1016/0095-8522(48)90072-5

Kumar, N., H. Mittal, V. Parashar, S.S. Ray and J.C. Ngila. 2016. Efficient removal of rhodamine 6G dye from aqueous solution using nickel sulphide incorporated polyacrylamide grafted gum karaya bionanocomposite hydrogel. *RSC Adv.* 6: 21929–21939.

Kumbhare, V. and A. Bhargava. 1999. Studies on the nutritional composition of *Sterculia* species. *J. Food Sci. Technol.* 36(6): 542–544.

Kuruwanshi, V.B., P. Katiyar and S. Khan. 2017. Scientific approaches of gum tapping in Gum Karaya (*Sterculia urens* Roxb.) for high gum production. *Int. J. Curr. Microbiol. App. Sci* 6(8): 3366–3374. https://doi.org/10.20546/ijcmas.2017.608.402

Le Cerf, D., F. Irinei and G. Muller. 1990. Solution properties of gum exudates from *Sterculia urens* (Karaya gum). *Carbohyd. Polym.* 13, 375–386. https://doi.org/10.1016/0144-8617(90)90037-S

Lowry, J. 1971. Anthocyanins of some Malaysian *Sterculia* species. *Phytochemistry* 10: 689–690.

Lyzu, C., S. Mitra, K. Perveen, Z. Khan, A.M. Tareq, N.A. Bukhari, F.M. Husain, E.P. Lipy, D. Islam, M. Hakim, T.B. Emran and M.G. Dashti. 2022. Phytochemical profiling, antioxidant activity, and *in silico* analyses of *Sterculia villosa* and *Vernonia patula*. *Evid. Based Complement. Alternat. Med.* 2022: 3190496. doi: 10.1155/2022/3190496.

Mangas, M.B.P., F.N. Rocha, P.A.Z. Suarez, S.M.P. Meneghetti, D.C. Barbosa, R.B. dos Santos, S.H.V. Carvalho and J.I. Soletti. 2012. Characterization of biodiesel and bio-oil from *Sterculia striata* (chicha) oil. *Industrial Crops Prod.* 36(1): 349–354. doi: 10.1016/j.indcrop.2011.10.021

Mayes, J.M. 2010. Gum Tragacanth and Karaya. In: *Food stabilisers, thickeners and gelling agents*. Alan Imeson Blackwell Publishing Ltd., London, pp. 167–179. doi:10.1002/9781444314724.ch9

Meer, W. 1980 Gum Karaya. In: R.L. Davidson (ed.), *Handbook of water soluble gums and resins*. McGraw-Hill, New York, pp. 10.1–10.14.

Mesaik, M.A., S. Rahat, K.M. Khan, M.I. Choudhary, S. Murad, Z. Ismail and A. Ahmad. 2004. Synthesis and immunomodulatory properties of selected oxazolone derivatives. *Bioorg. Med. Chem.* 12: 2049–2057. https://doi.org/10.1016/j.bmc.2004.02.034

Mikuš, Ľ., Ľ. Valík and L. Dodok. 2014 Usage of hydrocolloids in cereal technology. *Acta Univ. Agric. Silvic. Mendel Brun.* 59: 325–334. https://doi.org/10.11118/actaun201159050325

Miralles, J., E. Bassene and E.M. Gaydou. 1993. Determination of cyclopropenoid fatty acids in *Sterculia* seed oils from Senegal. *J. Am. Oil Chem. Soc.* 70: 205–206. http://dx.doi.org/10.1007/BF02542627.

Mujumdar, A.M., D.G. Naik, R.J. Waghole, D.K. Kulkarni and M.S. Kumbhojkar. 2000. Pharmacological studies on *Sterculia foetida* leaves. *Pharm. Biol.* 38: 13–17.

Nanadagopalan, V., M.J. Gritto and A. Doss. 2015. GC-MS analysis of biomolecules on the leaves extract of *Sterculia urens* Roxb. *J. Pharmacogn. Phytochem.* 3: 193–196.

Nitao, D.Z., B.P. Singh, N.V. Queiroga and C.J.M. Oliveira. 2008. *Sterculia striata* seed kernel oil: characterization and thermal stability. *Grasas Aceites* 59: 68–73.

Ohloff, G. 1978. Recent developments in the field of naturally-occurring aroma components. In: Herz, W., Grisebach, H., Kirby, G.W. (Eds.), Fortschritte der Chemie Organischer Naturstoffe/*Progress in the chemistry of organic natural products*. Springer-Verlag, Wien, pp. 431–527. https://doi.org/10.1007/978-3-7091-8505-6-5

Orisakeye, O.T. and T.A. Olugbade. 2014. Epicatechin and procyanidin B2 in the stem and root bark of *Sterculia tragacantha* Lindl (Sterculiaceae). *Med. Chem.* 4: 334–337.

Padil, V.V.T., S. Wacławek, M.C. Rajender and S. Varma. 2018. Tree gum-based renewable materials: Sustainable applications in nanotechnology, biomedical and environmental fields. *Biotechnol. Adv.* 36(7): 1984–2016. doi: 10.1016/j.biotechadv.2018.08.008

Padil, V.V.T., S. Wacławek, C. Senan, J. Kupčík, K. Pešková, M. Černík and H.M. Somashekarappa. 2017. Gum karaya (*Sterculia urens*) stabilized zero-valent iron nanoparticles: characterization and applications for the removal of chromium and volatile organic pollutants from water. *RSC Adv.* 7: 13997–14009.

Pawlowsky, N.E., J.E. Nixon and R.O. Sinnhuber. 1972. Assay of cyclopropenoid lipids by nuclear magnetic resonance. *J. Am. Oil Chem. Soc.* 49: 387–392.

Pham, N.K.T., T.D. Nguyen, T.D.C. Doan, T.D. Ha, N.M.A. Tran, T.D. Tran, D.T. Mai and T.P. Nguyen. 2021. Stercufoetin A, new oleanane-type triterpenoid from the leaves of *Sterculia foetida* L. *Nat. Prod. Res.* 35(7): 1226–1231. doi:10.1080/14786419.2019.1644508

Pham, N.K.T., T.Q.T. Nguyen, T.H. Nguyen, C.D. Huynh, B.L.C. Huynh, T.A.H. Tran, T.T.T. Nguyen, T.C.T. Dang and T.H.T. Nguyen. 2022. Phenolics and triterpenoids from the bark of *Sterculia foetida* Linn. *Viet Nam J. Chem.* 1: 70–75. https://onlinelibrary.wiley.com/doi/epdf/10.1002/vjch.202100081

Phillips, G.O. and P.A. Williams. 2001. Tree exudates gums: Natural and versatile food additives and ingredients. *Food Ingredients Analytical Int.* 23: 26–28.

Prastiwi, R., B. Elya, R. Sauriasari, M. Hanafi and E. Dewanti. 2018. Pharmacognosy, phytochemical study and antioxidant activity of *Sterculia rubiginosa* Zoll. ex Miq leaves. *Pharmacog. J.* 10(3): 571–575. doi: 10.5530/pj.2018.3.93

Rao, P. and R. Sharma. 1957. Studies on Indian plant gums: composition and graded hydrolysis of gum karaya (*Sterculia urens* Roxb.). *Proc Indian As-Math Sci* 45: 24–29 *https://doi.org/10.1007/BF03052571*

Sahu, P., P.S. Pisalkar, S. Patel and P. Katiyar. 2019. Physico-chemical and rheological properties of karaya gum (*Sterculia urens* Roxb.). *Int. J. Curr. Microbiol. App. Sci.* 8(4): 672–681. https://doi.org/10.20546/ijcmas.2019.804.072

Sao, K.P. 2012. Physico-chemical properties of some Indian plant gums of commercial importance. Indian Institute of Natural Resins and Gums, Indian Council of Agricultural Research, Namkun, Ranchi (Jharkhand). *Tech. Bullet.* 1: 1–64 *https://iinrg.icar.gov.in/library/physico/mobile/index.html#p=15*

Sarah, L.V., B.W. Singer, S.M. Hicchen and J.H. Townsend. 1998. The development and initial application of gas chromatographic method for the characterization of gum media. *J. Amer. Inst. Convers.* 37: 295–311.

Shukla, A., K. Desai and N. Modi. 2020. In vitro antioxidant and antimicrobial potential of *Sterculia urens* Roxb. root extract and its bioactive phytoconstituents evaluation. *Futur J Pharm Sci* 6: 45. https://doi.org/10.1186/s43094-020-00063-9

Silva, D., A. Brito, R. De Paula, J. Feitosa H. Paula. 2003. Effect of mono and divalent salts on gelation of native, Na and deacetylated *Sterculia striata* and *Sterculia urens* polysaccharide gels. *Carbohyd. Polym.* 54: 229–236.

Slaghenaufi, D., M.C. Perello, S. Marchand-Marion, S. Tempere and G. de Revel. 2014. Quantitative solid phase microextraction–gas chromatography mass spectrometry analysis of five megastigmatrienone isomers in aged wine. *Anal. Chim. Acta* 813: 63–69. https://doi.org/10.1016/j.aca.2014.01.019

Slaghenaufi, D. and M. Ugliano. 2018. Norisoprenoids, sesquiterpenes and terpenoids content of valpolicella wines during aging: Investigating aroma potential in relationship to evolution of tobacco and balsamic aroma in aged wine. *Front Chem* 6: 66. https://doi.org/10.3389/fchem.2018.00066

Shrivastava, A. and V. Saxena. 1983. Study of nonsaponifiable principles from the root of *Sterculia foetida. Natl. Acad. Sci. Lett.* 6: 51–52.

Stephen, A.M., G.O. Phillips and P.A. Williams. 2006. *Food polysaccharides and their applications.* 2nd edition. CRC/Taylor & Francis, London.

Taher, A.A. 1998. Analytical and structural studies on some plant gum exudates. M.Sc. Thesis, Faculty of science, University of Kordofon.

Tandon, N. and M. Sharma. 2009. *Reviews on Indian Medicinal Plants.* Vol 8 (Da-Dy). Medicinal Council of Medical Research, New Delhi.

Thabet, A.A., F.S. Youssef, M. El-Shazly and A.N.B. Singab. 2018. *Sterculia* and *Brachychiton*: a comprehensive overview on their ethnopharmacology, biological activities, phytochemistry and the role of their gummy exudates in drug delivery. *J Pharm Pharmacol.* 70(4): 450–474. doi: 10.1111/jphp.12876.

Van Cott, H., C.E. Hamilton and A. Littell. 10 April 1954. The effects of subthreshold concentrations of monosodium glutamate on absolute thresholds. In: *Proceedings of the 75th Annual Meeting Eastern Psychological Association*. New York.

Verbeken, D., S. Dierckx and K. Dewettinck. 2003. Exudate gums: occurrence, production, and applications. *Appl. Microbiol. Biotechnol.* 63: 10–21. https://doi.org/10.1007/s00253-003-1354-z

Vickery, J.R. 1980. The fatty acid composition of seed oils from ten plant families with particular reference to cyclopropene and dihydrosterculic acids. *J. Am. Oil Chem. Soc.* 57: 87–91.

Vinod, V.T.P., S. Chandra and C. Miroslav. 2015. Dodecenylsuccinic anhydride derivatives of gum karaya (*Sterculia urens*): Preparation, characterization, and their antibacterial properties. *J. Agric. Food Chem.* 63: 3757 – 3765. doi: 10.1021/jf505783e

Vinod, V.T.P., R.B. Sashidhar, V.U.M. Sarma and S.S. Raju. 2010. Comparative amino acid and fatty acid compositions of edible gums kondagogu (*Cochlospermum gossypium*) and karaya (*Sterculia urens*). *Food Chem.* 123(1): 57–62.

Wareing, M.V. 1997. Exudate gums. In: A. Imeson (ed.), *Thickening and gelling agents for food*. 2nd edition. Blackie Academic & Professional, London, pp. 86–118.

Wang, R.F., X.W. Wu and D. Geng. 2013. Two cerebrosides isolated from the seeds of *Sterculia lychnophora* and their neuroprotective effect. *Molecules* 18(1): 1181–1187.

Wang, R.F., X.W. Yang, C.M. Ma, M.Y. Shang, J.Y. Liang, X. Wang, S.Q. Cai and Y. Shoyama. 2003. Alkaloids from the seeds of *Sterculia lychnophora* (Pangdahai). *Phytochemistry* 63: 475–478.

Wang, W. 2000. Tragacanth and karaya. In: G.O. Phillips and P.A. Williams (eds), *Handbook of hydrocolloids*. CPC Press LLC, New York, pp. 231–246.

Weiping, W. and A. Branwell. 2000. Tragacanth and karaya. In: G.O. Phillips and P.A. Williams (eds), *Handbook of hydrocolloids*. Woodhead Publishing Limited, New York pp. 231–245.

Whistler, L.R. and N.J. BeMiller. 1993. *Industrial gums: Polysaccharides and their derivatives*. 3rd edition. Academic Press, Inc., San Diego.

Wu, Y., L. Ai, J. Wu and S.W. Cui. 2013. Structural analysis of a pectic polysaccharide from boat-fruited *Sterculia* seeds. *Int. J. Biol. Macromol.* 56: 76–82. doi: 10.1016/j.ijbiomac.2013.01.005.

Xia, P.F., Z.M. Feng, Y.N. Yang and P.C. Zhang. 2009b. Two flavonoid glycosides and a phenyl propanoid glucose ester from the leaves of *Sterculia foetida*. *J. Asian Nat. Prod. Res.* 11(8): 766–71. doi: 10.1080/10286020903055103.

Xia, P.F., S. Song, Z. Feng and P. Zhang. 2009a. Chemical constituents from leaves of *Sterculia foetida*. *Zhongguo Zhong Yao Za Zhi (=China journal of Chinese materia medica)* 34(20): 2604–2606.

Yadav, S., P.K. Sharma, K. Goyal and Narendra. 2015. Comparative study of mucilage extracted from seeds of *Cassia fistula* and gum karaya. *Adv. Biological Res.* 9(3): 177–181.

Yesodharan, K. and K.A. Sujana. 2007. Wild edible plants traditionally used by the tribes in the Parambikulam wild life sanctuary, Kerala, India. *Nat. Prod. Rad.* 6(1): 74–80.

Yoshiki, Y. and K. Okubo. 1995. Active oxygen scavenging activity of DDMP (2, 3-dihydro-2, 5-dihydroxy-6-methyl-4 H-pyran-4-one) saponin in soybean seed. *Biosci. Biotechnol. Biochem.* 59: 1556–1557. https://doi.org/10.1271/bbb.59.1556

Zeomar, N.D., S.B. Pushkar, Q.N. Vicente and O.C.J. Marcelino. 2008. *Sterculia striata* seed kernel oil: characterization and thermal stability. *Grasasy Aceites* 59(2): 160–165.

6 Traditional Uses, Pharmacology, and Industrial Applications of Gum Karaya

K. Venkata Ratnam, Lepakshi Md. Bhakshu, and T. Pullaiah

6.1 INTRODUCTION

Gum karaya is an exudate of the stem bark of *Sterculia urens* Roxb. and other species of *Sterculia. S. urens* is a large deciduous tree of tropical forests. Gum karaya is a commercially viable polysaccharide gum extracted from the *Sterculia urens* tree. The cacao tree (*S. setigera*) (Aspinall et al., 1965) and *S. caudata* (Hirst et al., 1958; Aspinall and Fraser, 1965) yield gums that, on partial hydrolysis, afford acidic oligosaccharides that resemble those obtained from *S. urens* gum.

Gum karaya is a negative colloid and a high-molecular-weight complex acidic polysaccharide. Gum karaya is chemically a partially acetylated polysaccharide of the substituted rhamnogalacturonoglycan (pectic) type. The composition of the gum is dependent on the species, and there is more than average variability in the proportions of amino acids in the proteinaceous components (Anderson et al., 1985b). Treatment of the gum of *Sterculia urens* with dilute alkali to saponify acetate groups has furnished a water-soluble and relatively homogeneous product (Aspinall and Nasir-ud-Din, 1965), of which about 60% of the sugar residues are neutral (Rha and Gal), and the rest acid (GalA and GlcA) (Aspinall and Sanderson, 1970). The general utility of gum karaya is based on its viscosity (Gershon and Pader, 1972). It was successfully evaluated for its suitability in the preparation of hydrophilic matrices (Babu et al., 2001a; Jain et al., 1992), mini-matrices (Cox et al., 1999), microcapsules (Babu et al., 2001b, 2001c), and transdermal patches (Mottaz et al., 1988).

Chemically, gum karaya is rich in polysaccharides, such as D-galacturonic acid, L-rhamnose, D-glucuronic acid, and D-galactose (Dikshith et al., 1984). It is extensively used in the pharmaceutical industry in different forms, such as powder, dental caps, hydrophilic matrices, hydrogels, floating tablets, floating beads, microcapsules, and microspheres to deliver drug at specific site. In the food industry, it is used as a stabilizer in sherbets and other preserved foods, to improve tenderness of biscuits, to give taste and odour in bread (Prasad et al., 2022). Scientifically, *S. urens* has been explored as an antibacterial, antifungal, antioxidant, cytotoxic, and lipid-lowering agent.

DOI: 10.1201/9781003438960-6

6.2 TRADITIONAL USES

6.2.1 *Sterculia africana*

Local tribes/traditional healers of Tanzania prescribe *S. africana* leaves to treat fungal infections (Moshi et al., 2006).

6.2.2 *Sterculia apetala*

Traditionally, *S. apetala* is used to treat skin diseases, coughs, malaria, and asthma. It is also used as a purgative and diuretic (Quattrocchi, 2012; Dvorak et al., 1998; Weeks and Black, 1992). It is also used to reduce weight, to promote glucose tolerance, and as an anti-inflammatory agent (Herrera-Meza et al., 2014).

6.2.3 *Sterculia coccinea*

African tribes and Chinese herbal practitioners use *S. coccinea* to treat gastrointestinal disorders. It shows potent activity in getting rid of bacterial infections (Mahbub et al., 2019).

6.2.4 *Sterculia foetida*

Sterculia foetida leaves have been reported as good insect repellent, diuretic, laxative, and antiepileptic in Indian traditional or folklore system (Mujumdar et al., 2000). The seed oil extracted from *S. foetida* is applied externally to treat skin diseases and itches. Seeds along with wood boiled and decoction are used to cure rheumatism (Raja et al., 2014).

6.2.5 *Sterculia guttata*

Juice obtained from the stem bark of *S. guttata*, along with the leaves of *Pogostemon benghalensis*, is given orally as a remedy for fever and diarrhoea. It is an important crop during famine. The seeds are eaten by local people during food scarcity (Katade et al., 2009).

6.2.6 *Sterculia lychnophora*

In Chinese traditional medicine, the dry seeds of *S. lychnophora* are a well-known and popular remedy for constipation (Wang et al., 2013). The seeds are also prescribed to treat sore throat and pharyngitis (Huang et al., 2012).

6.2.7 *Sterculia scaphigera*

The dry seeds of *S. scaphigera* are a popular remedy for constipation in Chinese traditional medicine (Feng, 1942).

6.2.8 *Sterculia setigera*

The seeds are eaten as purgative/dewormer. The oil from seeds has its use in local culinary and traditional medicine. The leaves and bark are of medicinal importance (Deshpande and Bhalsing, 2015). Oil from the seed is extracted on a local scale to be used in medicine (Rastogi and Mehrotra, 1993). *S. setigera* is used for the treatment of boils, inflammations, chickenpox, measles, dysentery, syphilis, epilepsy, jaundices, malaria, and leprosy traditionally (Igoli et al., 2005). The boiled leaves are used to treat malaria, and the stem bark decoction is used for the treatment of asthma, bronchitis, wound, fever, toothache, gingivitis sore, abscess, and diarrhoea (Lawal et al., 2010).

The fruits and root bark of *S. setigera* are used to treat dysentery, inflammation, measles, syphilis, chickenpox, epilepsy, malaria, boils, and leprosy. A decoction prepared from dry leaves of *S. setigera* is used to treat bacterial infections and chronic cough in AIDS patients and in blood strains (Babalola et al., 2012). The tribes of Sudan use *S. setigera* stem bark paste in external application to cure wounds. Water decoction prepared from dried stem bark of *S. setigera* is given orally in case of jaundice. A black powder prepared from dried seed and fruit mixture is applied externally to cure dermatitis by the Yoruba ethnic group of Nigeria. The Igedes tribes of Nigeria use stem bark decoction to cure diarrhoea (Tor-Anyiin et al., 2011).

6.2.9 *Sterculia striata*

Fresh leaves are made into a paste along with butter or olive oil, gently warmed, and applied externally to cure furuncles and boils. The plant parts are consumed by local tribes of Brazil (Silva et al., 2014; Sousa et al., 2012).

6.2.10 *Sterculia tragacantha*

African herbal practitioners prescribe a boiled decoction prepared from leaves, stem, and seeds of *S. tragacantha* used to alleviate arthritis, gout, diarrhoea, and oedema. The local people of the Nssuka region of Nigeria use *S. tragacantha* to reduce pain after bone fracture and also as a supporting material for orthopaedic patients (Udegbunam et al., 2013).

6.2.11 *Sterculia urens*

In Odisha, Kondhas, and Souras, the gum resin along with sugar is administered to cure dysentery (Pattanaik et al., 2006). In Tamil Nadu, the gum, along with honey, is used to treat throat problems (Sukumaran and Raj, 2010). Nandagoapalan et al. (2016) reported the therapeutic uses of stem bark as antidote to snakebites. In Andhra Pradesh, the stem bark of *Sterculia urens* is used to treat rheumatic pains and ulcers and gum for dysentery (Ravi Prasad Rao and Sunitha, 2011). Stem bark, mixed with *Haldina cordifolia* stem bark and black pepper, is boiled, and the decoction given orally for nine days to cure leucorrhoea (Venkata Ratnam and Venkata Raju, 2005). In Maharashtra, *S. urens* has been reported as remedy for nausea, skin diseases (Rupesh and Nitin, 2012), and arthritis (Kshirsagar, 2018). In Gujarat, *S. urens* stem bark, along with that of *S. villosa*, *Tecomella undulata*, and leaves of *Dalbergia volubilis*, is made into a paste and applied externally on bone fracture (Paul and Prajapati, 2014), while Riya et al. (2020) reported that fermented water extract of *S. urens* is applied externally to treat back pain. The local people of the Adilabad District, India, administer poultice prepared from the stem bark of *S. urens* and turmeric to cure rheumatoid arthritis (Suryanarayana Murthy and Seetharami Reddi, 2016).

In Bangladesh, pills prepared from gum karaya along with ginger, sulphur, and sugar crystals are given orally to remove stones in the urinary tract. Tablets prepared from gum karaya mixed with camphor, ginger, and sulphur are given orally to remove stones in the kidney/urinary tract (Shah et al., 2017). In Jharkhand, *S. urens* stem bark and gum are used to treat diarrhoea, stomach pain, and urinary tract infection (Singh, 2017). In the Satpura hill regions of Gujarat, the gum karaya is used as herbal remedy that acts as a laxative and fights throat infection (Patel and Varshney, 2017). The root of *S. urens*, along with milk, is given orally on empty stomach for 3–5 days to cure gonorrhoea by the Gond tribes of Madhya Pradesh (Ahmad and Fatima, 2018). The local tribes inhabiting the Gautam Buddha Wildlife Sanctuary of Bihar, India, administer *S. urens* root decoction to treat dysentery (Kumar, 2020). The local people of Toranmal Plateau of Maharashtra administer stem bark along with milk externally for bone fracture (Vilas, 2021). Fresh leaves along with coconut oil are made into a paste and applied on chronic wounds till complete healing (Sanjay, 2020).

Traditionally, gum karaya has been used to treat throat infections, to remove stones in the kidney/urinary tract, to treat diarrhoea, and as a laxative (Srivastava and Ray, 2015). In folk medicine, the stem bark and root parts of *S. urens* are used to treat bone fracture, menstrual problems, and gonorrhoea. The gum has been used to treat blood dysentery, joint pains, blisters, stomach problems, and throat infection (Luján-Medina et al., 2013, 2015). Gum mixed along with calcium carbonate is given orally for snakebite and applied externally on the bitten area. The gum extract is applied externally on the skin to remove spines, given orally to regularize menstrual problems (Jain et al., 2005).

Powdered gum fried along with jaggery and wheat flour is given orally for abdominal disorders and as nutritional food (Dhiman et al., 2019). The gum is also reported as remedy for bone fractures and burning sensation (Meena and Rao, 2010). Tamilselvi et al. (2019) reported that extract of gum karaya powder has significant effect on wound healing in diabetic patients.

S. urens leaves are used to cure pleuropneumonia in animals (Sukumaran and Raj, 2010). About 50 g of gum karaya, mixed with 20 g of jaggery, 10 g of bajra seeds, and turmeric, is made into a paste and applied externally on affected part to remove thorn in animals (Bhasker and Vinod, 2015).

6.2.12 *Sterculia villosa*

The Khampti and Lohit tribal communities of Arunachal Pradesh use *S. villosa* parts to treat inflammatory diseases, like cut wounds, burns due to hot water and fire, bone fracture, rheumatism, and muscle swelling (Namsa et al., 2009). The tribes of the Amara Kantak region of the Madhya Pradesh state use root paste as aphrodisiac and in erectile problems. Roots of *S. villosa*, along with an equal volume of fruits of *Glossogyne bidens*, *Peucedanum nagpurense*, *Cynoglossum lanceolatum*, and *Peucedanum dhana*, are made into a powder and given orally as an aphrodisiac (Kumar et al., 2004). Traditionally, *S. villosa* has been used as a therapeutic agent, as an aphrodisiac, as a diuretic, and in cooling. The cool drink prepared from petioles of *S. villosa* along with sugar and water is given orally to treat rheumatism and urinary problems. The stem bark and petiole are used as a potent remedy for seminal weakness. The sap of the tree stem bark is used as an effective medicine for throat infection. Root infusion is used as food supplement (Hossain et al., 2013).

S. villosa root and stem bark, along with other medicinal plant parts, like roots, stem bark, whole plant, rhizomes, and leaves, are used to treat bone fractures. *S. villosa* root is crushed and, after becoming a gel-like substance, applied as a bandage on bone fractures. *S. villosa* root, along with whole plant of *Tridax procumbens*, is made into a paste and applied as a bandage on bone fracture. Its root, along with that of *Grewia hirsuta*, rhizome of *Curcuma longa*, and tuberous root of *Dioscorea oppositifolia*, is made into a paste and applied on bone fractures. The stem barks of *S. villosa* and *S. urens*, along with that of *Tecomella undulata* and leaves of *Dalbergia volubilis*, are made into a paste and applied as a bandage on bone fractures (Paul and Prajapati, 2014). *S. villosa* stem bark is boiled in water, and the decoction given orally to get relief from tonsillitis (Sawmliana, 2013). Powdered root mixed with milk is given orally for easy delivery in pregnant women. Fresh or dry stem bark is made into a paste and applied on joints to get relief from joint pains. A small quantity of gum mixed with honey is given early in the morning to treat throat infection. Seeds ground into a paste along with jaggery are made into pills and given on an empty stomach to cure asthma and heart problems (Mali and Bhadane, 2008).

6.3 APPLICATIONS OF GUM KARAYA

Prasad et al. (2022) gave a detailed review on the applications of gm karaya. The applications of gum karaya are given in Figure 6.1 and Table 6.1.

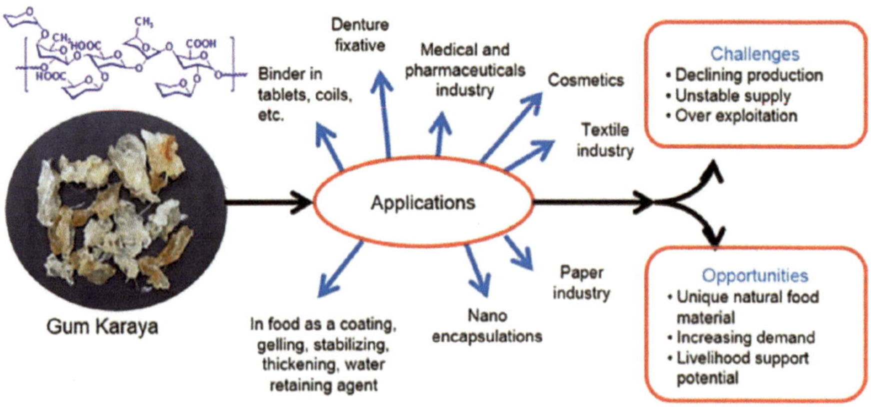

FIGURE 6.1 Applications of gum karaya.

Source: Prasad et al. (2022).

TABLE 6.1
Applications of gum karaya

Gum karaya applications	Specific uses	References
Medical application	Drug delivery	Waters et al., 1987
		Chavan et al., 2017
		Kumar et al., 2011
		Babu et al., 2002a, 2020d
		Postulkova et al., 2017
		Prajapati et al., 2013
		Freitas et al., 2020
		Bera et al., 2015
		Singh and Sharma, 2008, 2011
		Singh and Vashishtha, 2008
		Singh and Singh, 2018
		Singh et al., 2011
	Development of floating drug delivery system (FDDS)	Sarathchandiran and Suresh Kumar, 2013
		Singh et al., 2010
	Microspheres development for drug delivery at specific site	Sirisha et al., 2017
		Sreekanth Reddy et al., 2019
	Microencapsulation of drug	Kumar et al., 2011
		Luján-Medina et al., 2015
		Singh et al., 2010
		Singh and Sharma, 2009
		Upadhyay, 2017
	Bulk laxative	Munday and Cox, 2000
	Denture fixative	Steinhardt and Goldwater, 1962
		Jayaramudu et al., 2013
		Boiteau and Blondeel, 1981
	Nanoparticle synthesis	Padil and Černík, 2013
		Magalhães et al., 2016
	Wound plaster	Kaur et al., 2014
		Mirhosseini and Amid, 2012
		Sawayanagi et al., 1983
	Wound dressing	Singh and Pal, 2008
		Singh et al., 2021
		Marsan, 1967
	Application in disposable and permanent ileostomy appliance	Sanderson, 1996
		Sparberg et al., 1966
Food application	Stabilizer	Partyka, 1963
		Dziezak, 1991
	Binder	Beach and Salisbury, 1969
		Weiping and Branwell, 2000
	Adhesive	Allen and McCaleb, 1938
	Staling reducer	Weiping and Branwell, 2000

(Continued)

TABLE 6.1 (*Continued*)
Applications of gum karaya

Gum karaya applications	Specific uses	References
Bioremediation application	Removal of toxic heavy metal of industrial waste and polluted water using GK-based nanomaterials	Arora and Ramawat, 2016
	GK-based hydrogel for application in removing heavy metal from effluents of mine	Fosso-Kankeu et al., 2016
Other industrial applications	Thickening agent for direct printing in textile industries	Kala, 2016 Verbeken et al., 2003
	Tanneries	Omkar et al., 2012
	Nanofiber	Patra et al., 2015
	Plant micropropagation medium	Singh and Kaur, 2011
	Paper industry, as binding agent	Whistler, 1993

Source: Prasad et al. (2022).

6.3.1 Pharmaceutical Industry

6.3.1.1 Laxative

In the pharmaceutical industry, gum karaya is largely used as a very effective laxative due to its water absorption, mucilage forming, and water-insoluble capacity.

6.3.1.2 Denture Fixative

The powdered gum dust is applied as a denature adhesive in dental plates. Because of its rapid swelling capacity, it helps in creating a tight fit of the dental plate on teeth (Srivastava and Ray, 2015).

6.3.1.3 Drug Delivery

Another important property of gum karaya is erosion mechanism, which helps in cent per cent drug release in pharmaceuticals (Munday and Cox, 2000). Cox et al. (1999) investigated the S(+)-ibuprofen release capacity of gum karaya in the form of a hydrophilic matrix prepared along with hydroxymethyl cellulose and other additives. It is revealed that gum karaya released drugs faster than did the other material used. A bilayer matrix tablet is prepared using gum samples, such as gum karaya, xanthan gum, and guar gum, along with ethyl cellulose, for zero-order release of glipizide (Raghavendra Rao et al., 2010).

A novel flurbiprofen–gum karaya formulation was prepared to release flurbiprofen in controlled oral release in in vitro and in vivo experimental models. The data supports the correlation between release of drug in in vitro and absorption of drug in in vivo (Babu et al., 2002b). The anti-inflammatory efficiency of paste prepared from gum karaya was assessed in skin inflammations. The formulation significantly reduced skin inflammation from fistula discharge and allowed healing (Gross and Irving, 1977). Bart et al. (1989) evaluated the efficacy of a novel delivery formulation of salicylic acid along with gum karaya patches to treat verruca vulgaris in a clinical experiment. It revealed that volunteers that used salicylic acid–gum karaya patches significantly reduced warts than did controls.

In an *in vivo* experiment, papilloma induced with papilloma virus in healthy rabbits revealed that a new transdermal drug prepared from salicylic acid and gum karaya matrix significantly delivered salicylic acid and reduced inflammatory conditions in rabbits (Mottaz et al., 1988). Park and Munday (2004) evaluated the mucoadhesive controlled niacin delivery capacity of natural gum samples, such as gum karaya, xanthan gum, gaur gum, and glycol chitosan, in buccal drug delivery. Gum karaya exhibited superior adhesion capacity and cent per cent drug delivery capacity. A soft polymeric denticap was developed with gum karaya powder, ethyl cellulose, antibiotic, and analgesic to treat toothache and dental infections (Mukherjee et al., 2009). Hydrophilic matrices containing gum karaya and guar gum were developed to control the oral release of theophylline drug (Babu et al., 2003). The matrices extended the release of drug up to 24 hours.

The floating matrix tablets were prepared using natural polymers, such as gum karaya, guar gum, along with metoclopramide hydrochloride, and in combination with citric acid and calcium carbonate, to improve drug bioavailability (Singh et al., 2007). Senapati et al. (2006) developed phenylpropanolamine matrix tablets fabricated with gum karaya and guar gum alone and in combination with other excipients for the sustained release of drug (Figure 6.2). Gum karaya tablets demonstrated greater swelling capacity than guar gum. The tablets prepared with both gum samples demonstrated more sustained drug release than did the tables with individual

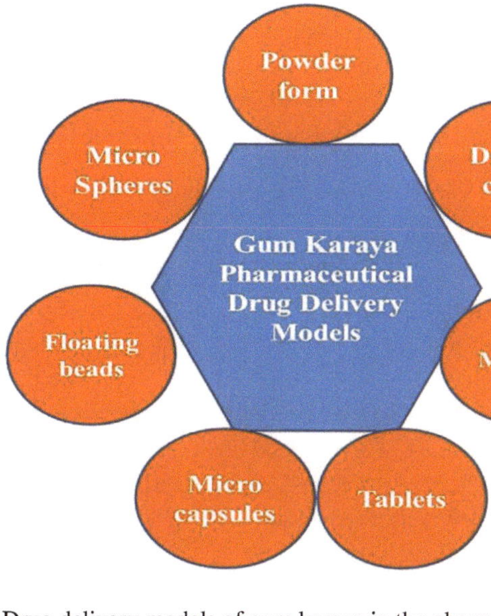

FIGURE 6.2 Drug delivery models of gum karaya in the pharmaceutical industry.

gum sample. Mini-matrices containing gum karaya, locust bean, and xanthan were examined to study the kinetics of diclofenac release in in vitro studies. It revealed that gum karaya exhibited the lowest drug release rate due to its moderate solubility and slow erosion rate (Sujja-areevatha et al., 1998).

A novel dental model was developed using gum karaya powder, Carbopol 934 P, poloxamer 407, and corn zein using the solvent evaporation technique to study the drug release capacity (Figure 6.2). The study results revealed that the drugs were stable in the formulation and released the drug at least for 24 hours (Ghosh et al., 2009). Singh and Sharma (2008) prepared glipizide microcapsules fabricated with gum karaya and alginate to evaluate drug release, mucoadhesive property, and to study *in vivo* hypoglycaemic activities. The capsules demonstrated good mucoadhesive property and helped in potent hypoglycaemic activity of glipizide. Similarly, gum karaya–based hydrogels were prepared for the effective delivery of antidiarrheal drug ornidazole (Singh and Sharma, 2011). The hydrogels played a dual role, that is, very good swelling property and cent per cent drug release mechanism in the treatment of diarrhoea.

Gastro-protective floating beads were prepared using gum karaya, alginate, and $CaCl_2$ as cross-linker to deliver antiulcer drug pantoprazole. The beads delivered the drug effectively and retained the drug for a long time (Singh et al., 2010). Sirisha et al. (2017) formulated microspheres using gum karaya for the site-specific delivery of cromolyn used to treat ulcerative colitis. Microspheres released the drug at a minimal level in in vitro conditions, which indicates the stability of drug in formulation. Sreekanth Reddy et al. (2019) developed a fabricated blend microspheres of gum karaya along with gelatin for the controlled release of distigmine bromide. Alange et al. (2017) developed pH-sensitive gum karaya microspheres to deliver colon-targeted anticancer drug. The microspheres reduced the release of drug in the upper portion of the intestine and enhanced drug release in the chronic region. Singh and Pal (2008) developed a novel wound dressing material with gum karaya to deliver the antibacterial drug tetracycline hydrochloride. Root juice mixed with honey is given orally to cure blood dysentery (Das et al., 2008).

6.3.2 Pharmacology of *Sterculia* Species and Karaya Gum

6.3.2.1 Antibacterial Studies

Antibacterial efficiency of esterified gum karaya revealed that the gum significantly controlled the growth of *Staphylococcus aureus*. These derivatives are promising in biomedical applications to control infectious diseases caused by *S. aureus* (Silva et al., 2020). Antibacterial activity of *S. urens* root hydro-methanolic extract showed that the extract exhibited various degrees of inhibition against the tested pathogens, such as *Klebsiella pneumoniae*, *E. coli*, and *Penicillium glaucum* (Shukla et al., 2020). The antibacterial effect of novel hydrogel prepared from gum karaya demonstrated that the gel strongly inhibited gram-positive bacterial strains than gram-negative strains (Lipový et al., 2018).

Karaya gum was investigated in vitro for its ability to prevent bacterial adhesion to denture acrylic. A protective coating of karaya gum applied to dentures reduced bacterial adhesion by 98% (Wilson, 1989). In an in vitro study, all microbial strains tested (*Staphylococcus aureus*, *Klebsiella pneumoniae*, *Candida albicans*, *Pseudomonas aeruginosa*) were completely or substantially inhibited by *S. urens* at a concentration of 1% and substantially or weakly inhibited at a concentration of 0.5% (Lipový, 2018). Antimicrobial activity against human pathogens was demonstrated by lectins extracted from the seeds of *S. foetida* (Braga, 2014).

Silver nanoparticles characterized from gum karaya along with polyvinyl alcohol exhibited potent antibacterial and bactericidal activity against the tested both gram-positive and gram-negative bacterial strains. Electrospun nanomembranes prepared from these nanoparticles may have many applications in the fields of medicine, food, and water purification (Vinod et al., 2015a). The antibacterial efficiency of esterified and deacetylated gum karaya samples is demonstrated against gram-negative and gram-positive bacterial strains (Vinod et al., 2015b). Copper oxide nanoparticles prepared using gum karaya demonstrated a broad spectrum of antibacterial activity against the tested bacterial strains (Padil and Černík, 2013). Antibacterial activity of ethyl acetate, hexane, and dichloromethane extracts of *S. setigera* leaf was evaluated against a virulent strain of *Mycobacterium tuberculosis*. It revealed that dichloromethane extract strongly inhibited the tested virulent strain (Babalola et al., 2012). Water extract of *S. setigera* was tested against five bacterial strains, namely, *Staphylococcus aureus*, *B. cereus*, *E. coli*, *Salmonella typhimurium*, and *Proteus mirabilis*, and indicated that *B. cereus* was strongly inhibited than other organisms (Konaté et al., 2011).

Antibacterial activity of ethanol extract of *S. foetida* leaves demonstrated concentration-dependant antibacterial activity against the tested pathogens, namely, *Staphylococcus aureus* and *E. coli* (Vital et al., 2010). Khatoon et al. (2016) evaluated the antibacterial activity of methanol and hexane extracts of *S. foetida* stem bark against six human pathogenic bacterial strains causing dysentery and diarrhoea using agar well diffusion method. The results indicated that *Shigella flexneri* was strongly inhibited by hexane extract, while methanol extract strongly inhibited *Salmonella enterica ser typhi*. Except *Klebsiella pneumoniae* and *Shigella flexneri*, the remaining bacterial strains exhibited moderate antibacterial activity to the tested extracts. Wilson et al. (2014) studied the antibacterial activity of the methanolic extract of *S. quinqueloba* leaves against five bacterial species and one yeast using micro-dilution method. It demonstrated that the extract strongly inhibited the tested pathogens.

Acetone, methanol, and petroleum ether extracts obtained from *S. setigera* were evaluated for antibacterial activity using disc diffusion and dilution methods, which revealed that all the tested extracts exhibited moderate antibacterial activity against the tested pathogens (Tor-Anyiin et al., 2011). Antibacterial activity of crude methanol and fractions of *S. villosa* bark demonstrated that all the tested organisms were inhibited moderately (Haque et al., 2014). *Salmonella paratyphi* was strongly inhibited by crude ethanol extract of *S. villosa* leaf (Tania et al., 2013). The inhibitory effects of ethanol extract of *S. lychnophora* were assessed against *Streptococcus mutans*, a pathogen of dental caries and plaque formation, and indicated that the extract strongly inhibited the growth of bacteria and reduced biofilm formation (Yang et al., 2016).

Antibacterial activity of silver nanoparticles synthesized from water and ethanol extracts of *S. diversifolia* demonstrated that the particles significantly inhibited the tested bacterial species *E. coli* and *S. aureus* (Al-Ramamneh et al., 2022).

6.3.2.2 Antifungal Activity

The antifungal activity of *S. urens* leaf aqueous, acetone, petroleum, ether, and chloroform extracts was tested using two methods, namely, poison food and agar well diffusion methods. It revealed that the poison food technique exhibited better results than the agar well diffusion method (Kanoje et al., 2021). Kanoje and Kavita (2017) reported very feeble antifungal activity of *S. urens* leaf water, methanol, hexane, and petroleum ether extracts against the tested clinically isolated fungal pathogens. Crude methanol and fractions of *S. villosa* bark demonstrated moderate antifungal activity (Haque et al., 2014).

The antifungal efficiency of dichloromethane, ethyl acetate, hexane, and butanol fractions of *S. setigera* leaves was investigated against four *Candida* species and four filamentous fungi using MIC and MFC methods. Among the tested pathogens, *Candida albicans* and *Aspergillus flavus* were strongly inhibited by ethyl acetate and dichloromethane fractions, respectively, in MIC method. While *Aspergillus flavus* and *A. niger* were significantly inhibited cent per cent by ethyl acetate and dichloromethane fractions in the MFC method (Ouédraogo et al., 2013). Ethanol crude extract of *S. villosa* leaf strongly inhibited *Pityrosporium ovale* (Tania et al., 2013). Methanolic water extract and isolated principal 3-hydroxyoctadecanoic acid showed significant antifungal activity against *Candida albicans* (Rollando et al., 2019; Ranta et al., 2012).

6.3.2.3 Antioxidant Activity

The DPPH scavenging capacity of *S. urens* root hydro-methanolic extract demonstrated that the extract strongly reduced DPPH purple at lower concentrations (Shukla et al., 2020). Palladium nanoparticles of gum karaya exhibited strong and concentration-dependent DPPH activity (Kumari et al., 2021). Hongsing et al. (2012) reported DPPH scavenging activity of gum karaya along with two gum samples, namely, *Acacia* and *Cochlospermum*. Antioxidant properties of a cerebroside isolated from the seeds of the related *S. lychnophora* plant were identified in an in vitro study; specifically, a moderate neuroprotective effect against SH-SY5Y cell damage induced by hydrogen peroxide was demonstrated (Wang, 2013). The antioxidant potential of the water extract of *S. setigera* was evaluated using FRAP and DPPH methods, demonstrating that the extract exhibited antioxidant activity in the FRAP method (Konaté et al., 2011).

The DPPH reducing capacity of hexane and methanol extracts of *S. foetida* stem bark demonstrated that both the tested extracts showed concentration-dependant DPPH scavenging activity. Hexane extract exhibited stronger DPPH scavenging activity than methanol extract (Khatoon et al., 2016). The antioxidant activity of methanol extract of *S. tragacantha* was assessed using FRAP and DPPH methods, demonstrating that the extract exhibited concentration-dependant FRAP and DPPH-mediated antioxidant activity (Udegbunam et al., 2011). Antioxidant activity of ethanol extract of *S. scaphigera* was evaluated (Daroi et al., 2013). DPPH scavenging activity of crude methanol and fractions of *S. villosa* bark strongly reduced DPPH purple colour in a concentration-dependant manner (Haque et al., 2014).

Water extract of *S. tragacantha* leaf demonstrated dose-dependent antioxidant activity in the tested models, such as ABTS, DPPH, FRAP, and total antioxidant capacity models (Onikanni et al., 2021). Antioxidant capacity of ethanol, methanol, and water extracts obtained from different parts, such as leaves, stem bark, and root parts, of *S. quadrifida* revealed that all the tested extracts exhibited significant DPPH scavenging activity (Amin et al., 2016; Dillak et al., 2019; Saefudin et al., 2013; Rollando and Monica, 2018; Lulan et al., 2018; Saragih and Siswadi, 2019).

6.3.2.4 Anthelmintic Activity

The anthelmintic efficiency of ethanol extract of *S. villosa* was evaluated against *Pheretima posthuma*. The results demonstrated concentration-dependant activity by inducing paralysis as well as death of worms (Alam et al., 2012).

6.3.2.5 Leishmanicidal Studies

Antileishmanial activities of ethanol extract of *S. villosa* stem bark were evaluated against *Leishmania donovani* promastigotes. It demonstrated that the extract strongly inhibited parasite at low concentrations (Das et al., 2017). Leishmanicidal activity of methanol crude extract and fractions like n-hexane, n-butanol, ethyl acetate, dichloromethane, and water extracts of *S. diversifolia* stem and leaf was evaluated in *Leishmania major* parasite. Among the tested ethyl acetate, a fraction of stem and leaf parts exhibited mild activity (Rabbi et al., 2020).

6.3.2.6 Diuretic Activity

The diuretic efficiency of ethanol extract of *S. villosa* was evaluated in rats and rabbit models. The results showed that the extract at 400 mg/kg body weight significantly increased urine output in both rat and rabbit models (Alam et al., 2012).

6.3.2.7 Anticonvulsant Activity

The anticonvulsant effect of ethanol extract of *S. foetida* was evaluated against MES- and pentylenetetrazole-induced (PTZ) convulsions in animal models. It revealed that the extract significantly altered MES-induced convulsions about 66%, whereas in the PTZ model, the extract delayed the onset of seizures (Raja et al., 2014).

6.3.2.8 Antinociceptive Activity

The antinociceptive efficiency of ethanol extract of *S. striata* stem bark against nociception induced by formalin, acetic acid, glutamate, and capsaicin demonstrated that the extract strongly reduced acetic acid–induced nociception, followed by formalin, capsaicin, and glutamate (Silva et al., 2014). Methanol extract of *S. tragacantha* demonstrated significant antinociceptive activity (Udegbunam et al., 2011).

6.3.2.9 Anti-Inflammatory Activity

Ethanol extract of *S. striata* stem bark significantly reduced dextran-induced oedema followed by carrageenan and histamine (Silva, 2014). The anti-inflammatory activity of ethanol extract of *S. villosa* was assessed in hypotonicity-induced cell membrane lysis in human red blood cells. The extract significantly protected the RBCs against hypotonicity-induced membrane lysis (Anwar et al., 2018). Wu et al. (2008) evaluated anti-inflammatory polysaccharides from seeds of *Sterculia* species. Ethanol extract of *S. foetida* leaves demonstrated significant anti-inflammatory activity in carrageenan-induced rat paw oedema (Mujumdar et al., 2000). Anti-inflammatory activity of methanol extract of *S. tragacantha* demonstrated that the extract significantly reduced carrageenan-induced paw oedema (Udegbunam et al., 2011). The anti-inflammatory activity of ethanol extract of *S. scaphigera* was evaluated (Daroi et al., 2013).

The anti-inflammatory efficiency of polysaccharides isolated from *Sterculia* seeds was assessed (Ai et al., 2012). Ethanolic extract of *S. villosa* stem bark significantly altered the carrageenan-induced paw oedema and cotton pellet–induced granuloma in albino rat models (Hossain et al., 2012). A bioactive phytoconstituent taraxer-14-en-3beta-ol, isolated from leaves of *S. foetida*, strongly inhibited TPA-induced ear oedema in mice models (Naik et al., 2004). Chloroform, n-hexane, ethyl acetate, and methanol extracts of *S. tragacantha* significantly reduced paw oedema and writhing time in animal models (Mogbojuri et al., 2016). Aqueous extract of *S. tragacantha* leaves strongly attenuated streptozotocin-induced higher levels of inflammatory markers, such as COX-2 and NO, in vivo models (Onikanni et al., 2021). Crude ethanol extract of *S. villosa* leaves exhibited moderate anti-inflammatory activity (Tania et al., 2013).

6.3.2.10 Antiatherothrombosis Activity

Antiatherothrombosis activity of crude ethanol extract obtained from *S. villosa* leaves was assessed using clot lysis method. It revealed that the extract showed moderate clot lysis in in vitro methods (Tania et al., 2013).

6.3.2.11 Antidiabetic Activity

Antidiabetic activity of ethanol extract of *S. villosa* was evaluated using α-amylase in in vitro conditions. The results demonstrate that the extract showed concentration-dependent α-amylase inhibition. It showed maximum inhibition at 1.5 mg/ml concentration (Anwar et al., 2018). Antidiabetic efficiency of methanolic extract of *S. foetida* leaves demonstrated that the extract significantly restored alloxan-induced metabolic symptoms in rat models. The antidiabetic activity of the extract was less potent than that of the standard drug glibenclamide (Shazia Hussain et al., 2014). Pre-treatment of ethanolic extract of *S. villosa* stem bark significantly altered alloxan-induced diabetic conditions in albino rat models (Hossain et al., 2012). In vitro antidiabetic effect of water extract of *S. tragacantha* leaves revealed that the extract significantly inhibited α-amylase and α-glucosidase enzymes, and it significantly restored streptozotocin-induced dysregulations in animal models (Onikanni et al., 2021).

6.3.2.12 Anticancer or Cytotoxic Activity

Antiproliferative properties of the bark of the related Madagascan *S. tavia* plant against a human ovarian cell line have been described (Dai, 2012). Cytotoxic efficiency of hexane, ethyl acetate, and dichloromethane extracts of *S. setigera* leaves was assessed against HepG2 cell lines and Vero cell lines, indicating that the extracts were non-toxic to the tested cell lines (Babalola et al., 2012). Cytotoxic study of esterified gum karaya against mammalian cells revealed that no cytotoxic effect was observed (Silva et al., 2020). Palladium nanoparticles synthesized using gum karaya demonstrated dose-dependent cytotoxic activity on MCF-7 cells (Kumari et al., 2021). Cytotoxic efficiency of methanol extract of *S. quinqueloba* against brine shrimp larvae demonstrated that the extract was non-toxic to the larvae (Wilson et al., 2014).

Antiproliferative activity of three phytoconstituents, namely, mansonone, tavinin, and epi-tavinin, isolated from endemic Madagascar plant *S. tavia* was investigated against ovarian cancer cell lines. It revealed that the three components exhibited moderate cytotoxic activity (Dai et al., 2012). Moshi et al. (2006) studied the cytotoxic activity of *S. africana* leaf methanol extract using brine shrimp larvae. The extract strongly supressed the growth of larvae. Cytotoxic activity of crude methanol and fractions of *S. villosa* bark was exhibited significantly on brine shrimp larvae (Haque et al., 2014). Cytotoxic potential of *S. diversifolia* stem bark methanol extracts and fractions against brine shrimp larvae and PC-3 cell lines showed concentration-dependent cytotoxicity and anticancer activity in the tested models (Rabbi et al., 2017).

Ethanol fraction extracted from *S. quadrifida* stem bark demonstrated moderate anticancer activity against T47D breast cancer cell lines (Rollando and Siswadi, 2016). Naptoquinone, namely, 2, 3-dihydro-6-hydroxy-2-methylenenaphtho[1,2-b] furan-4.5-dione, isolated from *S. quadrifida* stem bark strongly inhibited the proliferation of T47D cancer cells (Rollando, and Alfanaar, 2017). 2- iminoethyl 2-(2-(1-hydroxypentan-2-yl)phenyl) acetate isolated from ethyl acetate fraction and ethyl acetate fractions of *S. quadrifida* stem bark exhibited significant cytotoxic study against T47D cancer cells (Rollando et al., 2018 and Rollando and Prilianti, 2017). Ethyl acetate fraction of *S. quadrifida* stem bark demonstrated potent anticancer activity against T47D, MCF 7, and HepG2 cancer cell lines (Hertiani et al., 2019).

6.3.2.13 Anaesthetic Activity

Crude methanol extract and fractions obtained from *S. tragacantha* on local anaesthetic and effect on tissue was assessed using guinea pig models. It demonstrated that crude extract and fractions showed dose-dependent anaesthetic activity. Histological studies revealed no sign of tissue damage (Udegbunam et al., 2012). The anaesthetic potential of crude extract of *S. tragacantha* was investigated in goat models, revealing that the extract significantly reduced post-operative symptoms, such as blood glucose, pain threshold, and vital parameters (Udegbunam et al., 2013).

6.3.2.14 Analgesic Activity

Analgesic activity of crude methanol and petroleum ether fraction of *S. villosa* stem bark was evaluated in acetic acid–induced writhing in animal models. The results demonstrated that both extract and fraction significantly reduced acetic acid–induced pain (Hossain et al., 2013). Analgesic activity of ethanol extract of *S. scaphigera* was evaluated (Daroi et al., 2013).

6.3.2.15 Insecticidal or Antifeedant Activity

Insecticidal activity of acetone extract of *S. foetida* seeds was tested against two insects, namely, *Spodoptera litura* and *Achaea janata*, on castor plant using leaf disc method. The results indicated that seed extract showed a dual mode of action against the tested larvae, that is, it acted as insecticide against *S. litura* and as an antifeedant against *Achaea janata* (Rani and Rajasekharreddy, 2009). A similar type of study was undertaken with phytoconstituent, namely, (2n-octylcycloprop1-enyl)-octanoic acid (I) isolated from seeds of *S. foetida*, against three Coleopteran insects. It concluded that principal acted as a potent insecticide against the tested insects (Rani and Rajasekharreddy, 2010).

6.3.2.16 Depigmentation Activity

Inhibitory activity of the methanol extract of *S. lychanophera* on melanin synthesis in two types of melanoma cells, namely, melan-a cells and B16F10 melanoma cells, was assessed using in vitro methods. The results revealed that the extract strongly inhibited melanin synthesis in melan-a cells than in B16F10 cells. The inhibitory activity was confirmed with Western blot analysis (Lam et al., 2011).

6.3.2.17 Immunomodulatory Studies

The immunomodulatory activity of *S. diversifolia* stem bark methanol extracts and fractions was assessed using oxidative burst assay. The results demonstrated that ethyl acetate fraction strongly reduced the production of ROS, followed by dichloromethane and n-butanol (Rabbi et al., 2017).

6.3.2.18 Neuroprotective Activity

Neuroprotective efficiency of water extract of *S. tragacantha* leaf was investigated using acetylcholine esterase enzyme inhibition assay. It showed that the extract significantly inhibited acetylcholine esterase enzyme in in vitro conditions (Onikanni et al., 2021).

6.3.2.19 CNS Depressant Activity

CNS depressant activity of ethanol extract of *S. guttata* seeds in mice was evaluated. The results indicated that the extract showed dose-dependant CNS depressant activity (Katade et al., 2009). Ethanol extract of *S. foetida* leaves significantly reduced barbital sodium–induced CNS depression in mice model (Mujumdar et al., 2000).

6.3.2.20 Gastro-Protective Studies

Gastro-protective activity of ethanol of *S. striata* was assessed in ethanol- or ethanol/HCl-induced gastric lesions in mice models. The extract strongly inhibited ethanol-induced gastric lesions (Sousa et al., 2012).

6.3.2.21 Lipid-Lowering Activity

Cholesterol solubility efficiency of three gum samples, namely, gum karaya, acacia, and konda gogu, was evaluated in single-bile micelles (Hongsing et al., 2012). It indicated that gum karaya and acacia gum significantly decreased cholesterol solubility in bile micelles. Limited studies in hens and quails suggest karaya saponin may exert a hypocholesterolaemic effect (Afrose, 2010, 2011). No clinical data exist regarding the use of karaya extracts for the management of dyslipidaemia. Earlier preliminary studies evaluating effects of soluble fibres on plasma lipids, glucose tolerance, and mineral balance suggest that soluble refined gums, including karaya gum, may normalize blood sugar and plasma lipid levels (Behall, 1990).

Antihyperlipedaemic efficiency of methanolic extract of *S. foetida* leaves demonstrated that the extract significantly reduced alloxan-induced higher levels of liver enzymes and lipid profile in rat models. The antidiabetic activity of extract was less-potent than that of the standard drug glibenclamide (Shazia Hussain et al., 2014). The inhibitory effect of sterculic acid isolated from *S. foetida* seeds on stearoyl-CoA desaturase-1, a major regulator of lipid metabolism, was assessed in obese OLETF rats. Supplementation of seed oil demonstrated improved metabolic profile in obese rats. This may be due to the presence of cyclopropeonic fatty acids in the seed oil (Ortinau et al., 2013).

6.3.2.22 Lipase Inhibitory Activity

The inhibitory effect of gum karaya along with two gum samples, namely, *Acacia* and konda gogu, on lipase revealed that the gum samples exhibited concentration-dependent lipase inhibitory activity (Hongsing et al., 2012).

6.3.2.23 Larvicidal Studies

Larvicidal effect of bis(2-ethylhexyl) benzene-1,2-dicarboxylate isolated from seeds of *S. guttata* was assessed against two mosquito larvae, namely, *Culex quinquefasciatus* and *Aedes aegypti*, revealing that the constituent showed chronic toxicity on tested larvae (Katade et al., 2006a). The larvicidal effect of chloroform, ethanol, and hexane extracts of *S. guttata* seeds against two mosquito larvae demonstrated that all the extracts exhibited potent larvicidal activity (Katade et al., 2006b).

6.3.2.24 Tyrosinase Inhibition Study

The tyrosinase inhibitory effect of gum karaya, along with two gum samples, namely, *Acacia* and konda gogu, revealed that konda gogu showed better inhibition than the other two gum samples (Hongsing et al., 2012).

6.3.2.25 Laxative Property

Karaya gum particles absorb water and swell to 60 to 100 times their original volume, making it useful as a bulk laxative (Khan, 2009; Meier, 1990; Verbeken et al., 2003). Excessive doses as a laxative may cause diarrhoea and, with inadequate water consumption, may result in bowel obstruction (Duke, 1992). There are no recent animal data regarding the use of karaya gum as a laxative. Studies evaluating karaya gum laxative effects were conducted in dogs and rats in the 1930s (INCHEM, 2014).

6.3.2.26 Toxicity Studies

Sub-chronic toxicity concentrations of gum karaya studied in rat models revealed that the gum at 5% (w/w) reduced body weight in female rats (Taupin and Anderson, 1982). Toxicological studies of gum karaya have confirmed its status as a food additive (Meer, 1980; Glicksman, 1983; Eastwood et al., 1983). The gum is digested only to a very minor extent, if at all (Glicksman, 1983; Anderson et al., 1985a). The high solution viscosity of gum karaya, which, because of its acetate ester content (Goldstein and Alter, 1973), may be difficult to disperse, is a valuable characteristic (in whipped cream preparations, for example). The molecular structure of this complex, highly acidic pectic gum (of type B) has been investigated in considerable detail.

Acute oral toxicity of ethanolic extract of *S. villosa* stem bark in albino rat models demonstrated that the extract is safe up to 4,000 mg/kg body weight (Hossain et al., 2012). Water extract of *S. tragacantha* leaves showed no mortality up to 3,000 mg/kg body weight dose (Mogbojuri et al., 2016). Toxicity study of acetone extract of *S. villosa* seeds against two insects, namely, *Spodoptera litura* and *Achaea janata*, indicated that the extract at 2.5 mg/cm^2 exhibited significant toxicity against *Spodoptera litura* larvae (Rani and Rajasekharreddy, 2009).

6.3.2.27 Tablet Formulations

Babu et al. (2000) reported a modification to gum karaya (GK) to provide a lower viscosity and showed its suitability as a disintegrant in tablet formulations. Attempts were made to enhance nimodipine (NM) (insoluble in water) dissolution rate and efficiency by using modified gum karaya (MGK) as the hydrophilic carrier (Babu et al., 2002a). MGK was found to possess swelling comparable to that of the parent GK. In vivo evaluation of modified gum karaya as a carrier for improving the oral bioavailability of a poorly water-soluble drug, nimodipine (NM), was carried out by Babu et al. (2002c). The oral bioavailability was found to be NM powder < co-grinding mixtures of NM and gum karaya (GK) < co-grinding mixtures of NM and modified gum karaya (MGK) < NM solution. Authors concluded that MGK, an economical carrier, could be used for the dissolution enhancement of NM.

6.3.3 Industrial Applications of Gum Karaya

Gum karaya, an exudate from *S. urens*, has properties like gel- and film-forming, adhesiveness, water-absorbing, or moisture-absorbing abilities. Due to these abilities, the gum is used extensively in different industries, such as medicinal, food, cosmetic, pharmaceutical, and other industries, as meat binder, food stabilizer, bulk laxative, and denture powder (Goldstein, 1954). Raj et al. (2021) reviewed recent findings and future directions of grafted gum karaya polysaccharides and their various applications.

6.3.3.1 Food Industry

Gum karaya has multifarious applications in food industry due to its water-absorbing, water-holding, and acid compatibility. In the USA, it is used as a food stabilizer because it is listed as generally regarded safe (Ghodke et al., 2009). In France, gum karaya powder is used in sherbets, dressing, ice pops, and cheese spreads because of its ability to prevent 'bleeding' of water and large ice crystal formation (Srivastava and Ray, 2015). In the sugar industry, it improves the quality of juices, and it is a good emulsifier and binder in meat processing. In bakery products, it improves the appearance and tenderness of biscuits, taste, odour, cream colour, and softness of bread (Anupama et al., 2010). Karaya gum has been overlooked by the food industry for a long time; the natural, plant-based texturizer is both clean-label and functional, says French supplier Alland & Robert.

Gum karaya (GK) usage in the food industry includes 0.002% to 0.3% in meat, in dairy and bakery products, in flavours for flour confectionery, and exceptionally, 0.9% in American confectionery (FDA, 1974). GK is particularly useful at the 0.1–0.25% level in salad dressings, mayonnaise, ketchups, sauces, chutneys, cheese spreads, etc, where its excellent long-term acid stability gives products high quality, long shelf-life, and organoleptic effects that cannot be achieved with other hydrocolloids. The use of GK allows manufacturers to meet the increasing demand of a rapidly expanding segment of the market for 'all natural' products, in which synthetic or chemically modified stabilizers/thickeners cannot be used (Anderson, 1989).

Gum karaya, accepted as generally recognized as safe (GRAS) in the USA since 1961, was accepted temporarily (Annex II) as a food additive by the EEC in 1974. Since then, no adverse incident involving human health has been attributed to the ingestion of gum karaya, which is used in extremely small amounts in foods. The report by Anderson (1989) collates the evidence of safety now available and presents the data on need, production tonnages, and dietary intake levels.

Sterculia species seeds are used for human food. Kumbhare et al. (1999) investigated the nutritional composition of seeds of *Sterculia urens* and *S. foetida* from four localities. The results showed that carbohydrates ranged from 18 to 48%, crude protein from 9.63 to 30.81%, ascorbic acid from 2.3 to 5.2 mg/100g, and fat from 24 to 32.5%. Appreciable amounts of minerals were also present. Tannin content was found to be appreciably low.

6.3.3.2 Bioremediation

Natural or plant-based are good sources of polymer-based coagulants. Several researches identified that gum karaya is an ideal choice for natural polymer-based coagulants to remove heavy metals from soils. Vinod et al. (2011) reported that gum karaya is a good source of polymer-based coagulant to remove mercury ions from aqueous solutions at different temperature, pH, and contact time. Efficiency of the heavy metal removal capacity of nanoparticles and nanofibre membranes prepared from gum karaya was reviewed (Arora and Ramawat, 2014; Vinod et al., 2015b). Gum karaya–based iron nanoparticles were characterized to remove chromium and volatile water pollutants (Vinod et al., 2017). Gum karaya–based hydrogel was prepared using iron oxide to pull out heavy metals from mine effluents (Fosso-Kankeu et al., 2016).

Gum karaya–based hydrogels were prepared and grafted with 2-(dimethylamino) ethyl methacrylate and N, N-methylene-bis-acrylamide, which exhibited adsorption capacity of methylene blue and indigo caramine 89 to 101mg/g. At acidic pH, the gel exhibited desorption (Preetha and Vishalakshi, 2020). Mittal et al. (2020) studied the efficiency of the hydrogel of gum karaya grafted with acrylic acid and N-isopropylacrylamide for its effective removal capacity of cationic dyes from wastewater. It revealed that the hydrogel effectively removed brilliant green dye. Further, the nanofibres prepared using gum karaya have been used as biodegradable, biocompatible, and cost-effective materials in biomedical and material science fields (Patra et al., 2015).

6.3.4 Other Industrial Uses

In the textile industry, gum karaya is used as a thickening agent for dye in direct-colour printing operations (Abo-Shosha et al., 2008). In the paper industry, it is used as binder and deflocculated for the manufacture of lightweight paper. In the petroleum industry, it is applied as a thickening agent for drilling mud and in plugging vehicles in secondary recovery operations (Srivastava and Ray, 2015).

6.3.4.1 Applications in Oil Drilling

Metal-cross-linked polymer gels have been playing a vital role in stimulating oil and gas wells for many years, but the presence of large insoluble residues and the incomplete breaking of the conventional guar-based fracturing fluids reduce the permeability of the proppant pack necessary for oil and gas flow. Chauhan et al. (2019) synthesized a novel cross-linked gel using the low-molecular-weight acidic polymer gum karaya and zirconium cross-linker as an alternative to guar-based fracturing fluids to increase proppant pack conductivity with competitive rheological performances. The polymer and cross-linker concentrations were optimized, and the synthesized gel was studied for structure, morphology, and time-independent and time-dependent rheological tests under simulated conditions of pressure and high temperature. The smaller amount of residues left by the fluid after breaking with oxidative breakers and the higher regained permeability of the sand pack system compared to guar confirm its ability to reduce the proppant pack damage.

6.4 CONCLUSION

The diversified medicinal uses of gum karaya in folklore as well as traditional systems have been proved and appreciated by the scientific community through in vitro and in vivo systems. In addition, gum karaya has found use in industrial applications, in drug delivery system, for many medicines. Gum karaya also found application in environment wellness through its bioremediation effects. Hence, gum karaya possesses wide beneficial implications in traditional use and has a lot of commercial importance, with crore of rupees in turnover yearly. Hence, gum karaya has proved that its pharmaceutical importance might increase demand on natural resources and threat on the life of mature trees. Hence, it should be conserved and effectively used through sustainable management.

REFERENCES

Abo-Shosha, M.H., N.A. Ibrahim, E. Allam and E. El-Zairy. 2008. Preparation and characterization of polyacrylic acid/karaya gum and polyacrylic acid/tamarind seed gum adducts and utilization in textile printing. *Carbohydr. Polym.* 74: 241–249. https://doi.org/10.1016/j.carbpol.2008.02.011

Afrose, S., M.S. Hossain and H. Tsujii. 2010. Effect of dietary karaya saponin on serum and egg yolk cholesterol in laying hens. *Br. Poult. Sci.* 51(6): 797–804.

Afrose, S., M.S. Hossain and H. Tsujii. 2011. Hypocholesterolemic effect of karaya saponin in Japanese laying quails (*Coturnix coturnix japonica*). *J. Anim. Physiol. Anim. Nutr. (Berl.).* 95(6): 693–700.

Ahmad, L.Z. and K. Fatima. 2018. Ethno-gynaecological health disorders and their ethnomedicinal treatment by the Gond tribals of Raisen district of Madhya Pradesh-India. *Intern. J. Adv. Res. Sci. Engin.* 7(4): 1976–1988.

Ai, L., J. Wu, N. Che, Y. Wu and S.W. Cui. 2012. Extraction, partial characterization and bioactivity of polysaccharides from boat-fruited *Sterculia* seeds. *Int. J. Biol. Macromol.* 51(5): 815–818.

Alam, Md. R., Md. Raton, Md. M. Hassan, Md. F. Kadir and Md. A. Haque. 2012. Anthelmintic and diuretic activity of bark extracts of *Sterculia villosa. J. Appl. Pharmaceut. Sci.* 2(10): 86–89.

Alange, V.V., R.P. Birajdar and R.V. Kulkarni. 2017. Functionally modified polyacrylamide-graft-gum karaya pH-sensitive spray dried microspheres for colon targeting of an anti-cancer drug. *Intern. J. Biol. Macromolecules*, 102: 829–839.

Allen, H.E. and A.G. McCaleb. 1938. A method or process of manufacturing sausages and sausageflling composition. GB Patent application – 48.

Al-Ramamneh, E.M., A.M. Ghrair, A.K. Shakya, K.Y. Alsharafa, K. Al-Ismail, S.Y. Al-Qaraleh, J. Mojski and R.R. Naik. 2022. Efficacy of *Sterculia diversifolia* leaf extracts: Volatile compounds, antioxidant and anti-inflammatory activity, and green synthesis of potential antibacterial silver nanoparticles. *Plants*, 11: 2492.

Amin, A., J. Wunas and Y.M. Anin. 2016. Antioxidant activity test of klika Faloak ethanol extract (*Sterculia quadrifida* R. Br.) using DPPH (2,2-diphenyl-1-picrylhydrazyl). *Indonesian Fitofarmaka J.* 2(2): 111–114.

Anderson, A.W.J., W.J. Brydon, M.A. Eastwood, F.J. McDougall and D.M.W. Anderson. 1985a. Absence of rhamnose in human urine following ingestion of gum karaya. *Food Addit. Contam.* 2: 33.

Anderson, D.M.W. 1989. Evidence for the safety of gum karaya (*Sterculia* spp) as a food additive. *Food Addit. Contam.* 6: 189–199. https://doi.org/10.1080/02652038909373774

Anderson, D.M.W., J.F. Howlett and C.G.A. McNab. 1985b. Amino acid composition of proteinaceous component of gum karaya (*Sterculia* spp.). *Food Addit. Contam.* 2: 153.

Anupama, S., S. Goyal and N. Goyal. 2010. Applications of gum karaya in drug delivery systems: A review on recent research. *Der Pharmacia Lettre* 2(5): 39–48.

Anwar, R., S.R. Sultana, S. Hossen and Md. R.H. Chowdhury. 2018. Ethnopharmacological investigation in *Sterculia villosa* to determine anti-diabetic, anti-inflammatory, antioxidant, thrombolytic, and cytotoxic effect. *J. Pharmacogn. Phytochem.* 7(2): 3203–3211.

Arora, J. and K. Ramawat. 2014. Biology and biotechnology of gum yielding Indian trees. *Tree Biotechnol.* https://doi.org/10.1201/b16714

Aspinall, G.O. and R.N. Fraser. 1965. Plant gums of the genus *Sterculia*. Part II. *Sterculia caudata* gum. *J. Chem. Soc.* 4318.

Aspinall, G.O., R.N. Fraser and G.R. Sanderson. 1965. Plant gums of the genus *Sterculia*. Part III. *Sterculia setigera* and *Cochlospermum gossypium* gums. *J. Chem. Soc.* 4325

Aspinall, G.O. and Nasir-ud-Din. 1965. Plant gums of the genus *Sterculia*. Part I. The main structural features of *Sterculia urens* gum. *J. Chem. Soc.* 2710.

Aspinall, G.O. and G.R. Sanderson. 1970. Plant gums of the genus *Sterculia*. Part V. Degradation of carboxyl-reduced *Sterculia urens* gum, *J. Chem. Soc.* (C), 2259.

Babalola, I.T., E.A. Adelakun, Y. Wang and O.F. Shode. 2012. Anti-TB activity of *Sterculia setigera* Del, leaves (Sterculiaceae). *J. Pharmacogn. Phytochem.* 1(3): 17–23.

Babu, G.V., K. Himasankar, B. Janaki Ram, et al. 2002d. Studies on preparation and evaluation of modified form of gum karaya. *Indian J. Pharm. Sci.* 64: 244–249.

Babu, G.V., K. Himasankar, N.R. Kumar and K.V. Murthy. 2003. Influence of colon degradation of polysaccharide on the oral bioavailability of theophylline from controlled release hydrophilic matrices. *Boll. Chim. Farm.* 142(10): 454–60.

Babu, G.V.M.M., K. Himasankar, C.P.S. Narayan and K.V. Ramana Murthy. 2001a Evaluation of gum karaya in the design of oral controlled release hydrophilic matrix systems. *Acta Pharm.* 51: 273–287.

Babu, G.V.M.M., K. Himasankar, C.P.S. Narayan and K.V. Ramana Murthy. 2001c. Controlled release of diclofenac sodium by gum karaya–chitosan complex coacervate: In vivo evaluation. *Ind. J. Pharm. Sci.* 63: 408–412.

Babu, G.V.M.M., N.R. Kumar, K.H. Sankar, B.J. Ram, N.K. Kumar and K.V.R. Murthy. 2002a. In vivo evaluation of modified gum karaya as a carrier for improving the oral bioavailability of a poorly water-soluble drug, nimodipine. *AAPS PharmSciTech* 3(2): Article 12.

Babu, G.V.M.M., C.H.D.S. Prasad, C.P.S. Narayan and K.V. Ramana Murthy. 2001b. New system for microencapsulation of diclofenac sodium using gum karaya and chitosan. *Saudi Pharm. J.* 9: 168–179.

Babu, G.V.M.M., C.H.D.S. Prasad and K.V. Ramana Murthy. 2002c. Evaluation of modified gum karaya as carrier for the dissolution enhancement of poorly water-soluble drug nimodipine. *Int J Pharm.* 234: 1–17.

Babu, G.V.M.M., C.H.D.S. Prasad, N. Ravi Kumar and K.V. Ramana Murthy. 2000. Studies on the modified form of gum karaya and its applicability as tablet disintegrant. *Int. J. Pharm. Excip.* 2(2): 185–191.

Babu, G.V.M.M., D.S. Prasad, K. Himasankar, V. Gourisankar, N.K. Kumar and K.V.R. Murthy. 2002b. Development of new controlled release formulation of flurbiprofen: In vitro-in vivo correlation. *Indian J. Pharmaceut. Sci.* 64(1): 37–43.

Bart, B.J., J. Biglow, J.C. Vance and J.L. Neveaux. 1989. Salicylic acid in karaya gum patch as a treatment for verruca vulgaris. *J. Am. Acad. Dermatol.* 20(1): 74–76.

Beach, P.L. and M. Salisbury. 1969. Preparation of a proteinaceous food salad. *United States Patent* 28: 131–134.

Behall, K.M. 1990. Effect of soluble fibers on plasma lipids, glucose tolerance and mineral balance. *Adv. Exp. Med. Biol.* 270: 7–16.

Bera, H., S. Boddupalli and A.K. Nayak. 2015. Mucoadhesive-floating zinc-pectinate-sterculia gum inter-penetrating polymer network beads encapsulating ziprasidone HCl. *Carbohydr. Polym.* 131: 108–118.

Bhasker, P. and P. Vinod. 2015. Ethnoveterinary herbal practices used by tribes in Bhiloda (West) forest range, Aravalli District of Gujarat, India. *Indian J. Trad. Know.* 14(2): 313–318

Boiteau, C. and G. Blondeel. 1981. Tampon having a surface covering. GB Patient application 2,075,846

Braga, A.A.E., R.R. Lacerda, G.K. de Vasconcelos Medeiros et al. 2014. Antibacterial and hemolytic activity of a new lectin purified from the seeds of *Sterculia foetida* L. *Appl Biochem Biotechnol.* 175(3): 1689–1699.

Chauhan, G., A. Verma, A. Doley and K. Ojha. 2019. Rheological and breaking characteristics of Zr-crosslinked gum karaya gels for high-temperature hydraulic fracturing application. *J. Pet. Sci. Eng.* 172: 327–339. https://doi.org/10.1016/j.petrol.2018.09.043

Chavan, S.S., V.A. Pavlov and K.J. Tracey. 2017. Mechanisms and therapeutic relevance of neuro-immune communication. *Immunity.* 46: 927–942. https://doi.org/10.1016/j.immuni.2017.06.008

Cox, P.J., K.A. Khan, D.L. Munday and J. Sujja-areevath. 1999. Development and evaluation of a multiple unit oral sustained release dosage form for S(+)-ibuprofen: preparation and release kinetics. *Intern. J. Pharmaceut.* 193: 73–84

Dai, Y., L. Harinantenaina, P.J. Brodie, M.W. Callmander, S. Randrianasolo, E. Rakotobe, V.E. Rasamison and D.G. Kingston. 2012. Isolation and synthesis of two antiproliferative calamenene-type sesquiterpenoids from *Sterculia tavia* from the Madagascar rain forest. *Bioorg. Med. Chem.* 20(24): 6940–6944.

Daroi, P., S. Kasture and M. Mohan. 2013. Analgesic, anti-inflammatory, antioxidant and anti-ulcer activity of ethanolic extract of *Sterculia scaphigera* Hance (Sterculiaceae) seeds in mice and rats. *Int. J. Biol. Pharm. Res.* 4: 35–45.

Das, A., J.J. Jawed, M.C. Das, P. Sandhu, U.C. De, B. Dinda, Y. Akhter and S. Bhattacharjee. 2017. Antileishmanial and immunomodulatory activities of lupeol, a triterpene compound isolated from *Sterculia villosa. Int. J. Antimicrob. Agents*. 50(4): 512–522.

Das, A.K., B.K. Dutta and G.D. Sharma. 2008. Medicinal plants used by different tribes of Cachar district, Assam. *Indian J. Trad. Knowl.* 7(3): 446–454.

Deshpande, A. and S.R. Bhalsing. 2015. Sterculiaceae: A critical appraisal on plant tissue culture studies in medicinally important plants. *Res. Biotech.* 6(2): 31–38

Dhiman, M., A. Singh and M.M. Sharma. 2019. A review on *Sterculia urens* Roxb.: A boon to the livelihood for tribal people and industry. *Industrial Crops and Products*. 130: 341–351.

Dikshith, T.S.S., R.B. Raizada, R.B. Misra and K. Srivastava. 1984. Toxicological evaluation of karaya gum; acute and subacute oral toxicity in rats. *J. Biosci.* 6: 147–153.

Dillak, H.I., E.B.E. Kristiani and S. Kasmiyati. 2019. Secondary metabolites and antioxidant activity of ethanolic extract of Faloak (*Sterculia quadrifida*). *J. Biology & Biology Education*, 11(3): 296–303.

Duke, J.A. 1992. Handbook of biologically active phytochemicals and their activities. Boca Raton, FL: CRC Press Inc.

Dvorak, W.S., H. Urueña, L.A. Moreno and J. Goforth. 1998. Provenance and family variation in *Sterculia apetala* in Colombia. *Forest Ecology and Management*, 111(2/3): 127–135.

Dziezak, J.D. 1991. A focus on gums. *Food Technol.* 45: 116–132.

Eastwood, M.A., W.G. Brydon and D.M.W. Anderson. 1983. The effects of dietary gum karaya (*Sterculia*) in man. *Toxicol. Lett.* 17: 159–166. https://doi.org/10.1016/0378-4274(83)90052-8

FDA, Washington, 1974, GRAS re-affirmation of gum karaya. *Federal Register* 39(185): 34209–34211

Feng, D. 1942. *Sterculia scaphigera* (Pangtaihai or Tungtahai), a Chinese laxative drug. *J. Chin. Med.* 61: 9–11.

Fosso-Kankeu, E., H. Mittal and F. Waanders et al., 2016. Preparation and characterization of gum karaya hydrogel nanocomposite flocculant for metal ions removal from mine effluents. *Int. J. Environ. Sci. Technol.* 13: 711–724. https://doi.org/10.1007/s13762-015-0915-x

Freitas, A.A.R., A.J. Ribeiro and A.C. Santos et al., 2020. *Sterculia striata* gum as a potential oral delivery system for protein drugs. *Int. J. Biol. Macromol.* 164: 1683–1692. https://doi.org/10.1016/j.ijbiomac.2020.07.276

Gershon, S.D. and M. Pader. 1972. Dentrifices. In: Balsam, M.S. and E. Sagarin (eds.) Cosmetics science and technology. 2nd ed., Vol. 1. New York, NY: Wiley-Interscience, pp. 423–532

Ghodke., D.S., S.B. Patil, R.R. Shah, S.N. Nilofar, C.S. Magdum and D.D. Chougule. 2009. Evaluation of natural suspending properties from *Sterculia urens. Natural Prod.* 5(2): 54–56

Ghosh, S., G. Roy and B. Mukherjee. 2009. Dental mold: a novel formulation to treat common dental disorders. *AAPS Pharm. SciTech.* 10(2): 692–702

Glicksman. M. 1983. Gum karaya (*Sterculia* gum). In: Glicksman M (ed.), Food hydrocolloids, Vol. II. Boca Raton: CRC Press, p. 39

Goldstein, A.M. 1954. Chemistry, Properties, and Application of Gum Karaya. In: *Natural plant hydrocolloids*, Chapter 6. New York: American Chemical Society, pp. 33–37

Goldstein, A.M. and E.N. Alter 1973. Gum karaya. In: Whistler RL, BeMiller JN (eds.), Industrial gums, 2nd ed. New York: Academic Press.

Gross, E. and M. Irving. 1977. Protection of the skin around intestinal fistulas. *Br. J. Surg.* 64(4): 258–263.
Haque, S.S., M.M. Or Rashid, M.A. Md. Prodhan, S. Noor and A. Das. 2014. In-vitro evaluation of antimicrobial, cytotoxic and antioxidant activities of crude methanolic extract and other fractions of *Sterculia villosa* barks. *J. Appl. Pharm. Sci.* 4(3): 35–40.
Herrera-Meza, S., A.J. Martínez, M.G. Sánchez-Otero, M.R. Mendoza-López, O. García-Barradas and G.R. Ortiz-Viveros. 2014. Fatty acid composition and some physicochemical characteristics of *Sterculia apetala* seed oils. *Grasas y Aceites.* 65(3): e039. https://doi.org/10.3989/gya.0223141
Hertiani, T., Purwantiningsih, A. Winanta, W. Sasikirana, R. Munawaroh, E.P. Setyowati, R. Murwanti and Siswadi. 2019. *In vitro* immunomodulatory and cytotoxic potentials of faloak (*Sterculia quadrifida* R.Br.) bark. *OnLine J. Biol. Sci.* 19(4): 222–231.
Hirst, E.L., E. Percival and R.S. Williams. 1958. The structure of *Brachychiton diversifolium* gum (*Sterculia caudata*). *J. Chem. Soc.* 1942.
Hongsing, P., C. Palanuvej and N. Ruangrungsi. 2012 Chemical compositions and biological activities of selected exudate gums. *J. Chem. Pharmaceut. Res*, 4: 4174–4180.
Hossain, M.K., M.A. Prodhan, I.M.S.A. Hasan, H. Morshed and M.M. Hossain. 2012. Anti-inflammatory and antidiabetic activity of ethanolic extracts of *Sterculia villosa* barks on albino Wistar rats. *J. Appl. Pharm. Sci.* 2: 96–100.
Hossain, M.M., A.I.H. Even, M.A. Akbar, A. Ganguly and S.A. Rahman. 2013. Evaluation of analgesic activity of *Sterculia villosa* Roxb. (Sterculiaceae) bark in Swiss-Albino mice. *Dhaka University J. Pharmaceut. Sci.* 12(2): 125–129.
Huang, Y., T. Wu, L. Zeng and S. Li. 2012. Chinese medicinal herbs for sore throat. *Cochrane Database Syst. Rev.* 14 (3): CD004877. https://doi.org/10.1002/14651858.CD004877.
Igoli, J.O., O.G. Ogaji, T.A. Tor-Anyiin and N.P. Igoli. 2005. Medicinal practices among the Igede people of Nigeria Part II. *African J. Complem. Altern. Med.* 2 (2): 134–152.
INCHEM. 2014. Karaya Gum (Sterculia). International Programme in Chemical Safety INCHEM. www.inchem.org/documents/jecfa/jecmono/v024je06.htm.
Jain, A., S.S. Katewa, P.K. Galav and P. Sharma. 2005. Medicinal plant diversity of Sitamata wildlife sanctuary, Rajasthan, India. *J. Ethnopharmacol.* 102: 143–157.
Jain, N.K., K. Kulkarni and N. Talwar. 1992. Controlled-release tablet formulation of isoniazid. *Pharmazie.* 47(4): 277–278
Jayaramudu, J., G.S.M. Reddy, K. Varaprasad, et al., 2013. Preparation and properties of biodegradable flms from *Sterculia urens* short fber/cellulose green composites. *Carbohydr. Polym.* 93: 622–627. https://doi.org/10.1016/j.carbpol.2013.01.032
Kala, C.P. 2016. Important gum yielding species *Anogeissus latifolia* (Roxb.) Bed., *Boswellia serrata* Roxb. and *Sterculia urens* Roxb. ethnobotany, population density and management. *Appl. Ecol. Environ. Sci.* 4(3): 61–65. https://doi.org/10.12691/aees-4-3-2
Kanoje, T. and S. Kavita. 2017. Antifungal activity of some plant extracts against clinical pathogens. *Intern. J. Dev. Res.* 7(7): 14185–14188
Kanoje, T., K. Sharma and J.N. Verma. 2021. Qualitative phytochemical screening and in vitro antifungal activity of *Sterculia urens* Roxb. leaf extracts from Narayanpur, Bastar, Chhattisgarh. *Intern. J. Creative Res. Thoughts* 9(7): 195–212.
Katade, S.R., A.V. Misar, A.M. Mujumdar, U.D. Phalgune, R.D. Wakharkar and N.R. Deshpande. 2009. CNS depressant activity of ethanol extract of *Sterculia guttata* seeds in mice. *Indian J. Pharm. Sci.* 71(4): 445–447. https://doi.org/10.4103/0250-474X.57297.
Katade, S.R., P.V. Pawar, V.B. Tungikar, A.S. Tambe, K.M. Kalal, R.D. Wakharkar and N.R. Deshpande. 2006a. Larvicidal activity of bis(2-ethylhexyl) benzene-1,2-dicarboxylate from *Sterculia guttata* seeds against two mosquito species. *Chem. Biodivers.* 3(1): 49–53.
Katade, S.R., P.V. Pawar, R.D. Wakharkar and N.R. Deshpande. 2006b. *Sterculia guttata* seeds extractives–an effective mosquito larvicide. *Indian J. Exp. Biol.* 44(8): 662–665.

Kaur, M., T. Garg, G. Rath and A.K. Goyal. 2014. Current nanotechnological strategies for efective delivery of bioactive drug molecules in the treatment of tuberculosis. *Crit. Rev. Ther. Drug Carrier Syst.* 31(1): 49–88.

Khan, I. and E. Abourashed. 2009. Leung's encyclopedia of common natural ingredients used in food, drugs, and cosmetics. 3rd ed. Hoboken, NJ: Wiley.

Khatoon, A., A. Mohapatra and K.B.B. Satapathy. 2016. Studies on in vitro evaluation of antibacterial and antioxidant activities of *Sterculia foetida* L. Bark. *Int. J. Pharmaceu. Sci. Res.* 7(7): 2990–2995.

Konaté, K., M. Kiendrébéogo, M.B. Ouattara, A. Souza, A. Lamien-Meda, Y. Nongasida, N. Barro, J. Millogo-Rasolodimby and O.G. Nacoulma. 2011. Antibacterial potential of aqueous acetone extracts from five medicinal plants used traditionally to treat infectious diseases in Burkina Faso. *Current Res. J. Biol. Sci.* 5: 435–442.

Kshirsagar, S.R. 2018. Taxonomy and occurrence of some medicinal plants used in traditional medicines. *Arch. Pharm. Pharmacol. Res.* 1(3): 1–8.

Kumar, A. 2020. A contribution to the ethnomedicinal plants of Gautam Buddha Wildlife Sanctuary, Bihar & Jharkhand. *Int. J. Adv. Res. Biol. Sci.* 7(11): 108–115.

Kumar, A., T. Balakrishna and J. Rajiv. 2011. Formulation and evaluation of mucoadhesive microcapsules of metformin HCL with gum karaya. *Int. J. Pharm. Pharm. Sci.* 3: 150–155.

Kumar, R., N.R. Suman and S.S. Dash. 2004. Traditional uses of plants by tribals of Amara kantak region, Madhya Pradesh. *Indian J. Trad. Knowl.* 3: 383–390.

Kumari, A.S., M. Venkatesham and G. Veerabhadram. 2021. Monowave synthesis of Palladium nanoparticles by *Sterculia urens* gum (karaya) activity on MCF-7 Human Breast Cancer Cells and antioxidant activity. *Compliance Engineering J.* 12(6): 488–500.

Kumbhare, V. and A. Bhargava. 1999. Studies on the nutritional composition of *Sterculia* species. *J. Food Sci. Technol.* 36: 542–544.

Lam, U.D.P., D.N. Hoang, H.-B. Lee, B. Kim, J.-D. Lee, J.-H. Shin and E.-K. Kim. 2011. Depigmenting effect of *Sterculia lynchnophera* on B16F10 melanomaand C57BL/6 melan-a cells. *J. Chem. Eng.* 28(4): 1074–1077.

Lawal, I.O., N.E. Uzokwe and A.B. Igboanugo. 2010. Ethnomedicinal information on collation and identification of some medicinal plants in Research Institutes of South-west Nigeria. *African J. Pharma. Pharmacol.* 4(1): 1–7.

Lipový, B., J. Jholoubek, L. Vacek, F. Růžička, E. Nedomova, H. Poštulková and L. Vojtová. 2018. Antimicrobial effect of novel hydrogel matrix based on natural polysaccharide *Sterculia urens. Čs. epidemiologie, mikrobiologie, imunologie. Praha: ČSL JEP* 67(4): 166–174.

Luján-Medina, G.A., J. Ventura, J.A. Ascacio-Valdés et al., 2015. Microencapsulation of ellagic acid from pomegranate husk and karaya gum by spray drying. *Int. J. Pharm. Pharm. Sci.* 7: 212–216.

Luján-Medina, G.A., J. Ventura, A.C.L. Ceniceros, J.A. Ascacio, D.B.V. Valdes and C.N. Aguilar. 2013. Karaya gum: General topics and applications. *Macromolecules Indian J.* 9: 111–116.

Lulan, T.Y.K., S. Fatmawati, M. Santoso and T. Ersam. 2018. Antioxidant capacity of some selected medicinal plants in East Nusa Tenggara, Indonesia: The potential of *Sterculia quadrifida* R.Br. *Free Rad. Antioxid.* 8(2): 96–101.

Magalhães, G.A., E. Moura Neto, V.G. Sombra et al., 2016. Chitosan/*Sterculia striata* polysaccharides nanocomplex as a potential chloroquine drug release device. *Int. J. Biol. Macromol.* 88: 244–253. https://doi.org/10.1016/j.ijbiomac.2016.03.07

Mahbub, N., M. Nasrin Akter, T. Ahmed Shishir, S.M. Saiful Islam and N. Nawar. 2019. Pharmacological activity investigation of *Sterculia coccinea. J. Med. Plants Stud.* 7(4): 113–117.

Mali, P. and V. Bhadane. 2008. Some rare plants of ethnomedicinal properties from Jalgaon district of Maharashtra. *Intern. J. Green Pharmacy* 2(2): 76–78.

Marsan, A.E. 1967. Sealing pad for a post-surgical drainage pouch. US Patent application 3,302,647.

Meena, A.K. and M. Rao. 2010. Folk herbal medicines used by the Meena community in Rajasthan. *Asian J. Tradit. Med.* 5: 19–31.

Meer, W. 1980. Chap. 10. Gum karaya. In: Davidson, R.L. (ed.) *Handbook of water-soluble gums and resins*. New York: McGraw-Hill.

Meier, P., W.O. Seiler and H.B. Stahelin. 1990. Bulk-forming agents as laxatives in geriatric patients [in German]. *Schweiz Med Wochenschr.* 120(9): 314–317.

Mirhosseini, H. and B.T. Amid. 2012. A review study on chemical composition and molecular structure of newly plant gum exudates and seed gums. *Food Res. Int.* 46: 387–398. https://doi.org/10.1016/j.*foodres.2011.11.017*

Mittal, H., R. Babu, A.A. Dabbawala et al. 2020. Zeolite-Y incorporated karaya gum hydrogel composites for highly effective removal of cationic dyes. *Coll. Surf. A Physicochem. Eng. Asp.* https://doi.org/10.1016/j.colsurfa.2019.124161

Mogbojuri, O., A.A. Adedapo and M. Abatan. 2016. Phytochemical screening, safety evaluation, anti-inflammatory and analgesic studies of the leaf extracts of *Sterculia tragacantha. J. Complent. Integr. Med.* 13(3). https://doi.org/10.1515/jcim-2015-0114

Moshi, M.J., C.J. van den Beukel, O.J. Hamza, Z.H. Mbwambo, R.O. Nondo, P.J. Masimba, M.I. Matee, M.C. Kapingu, F. Mikx, P.E. Verweij and A.J. van der Ven. 2006. Brine shrimp toxicity evaluation of some Tanzanian plants used traditionally for the treatment of fungal infections. *Afr. J. Tradit. Complement. Altern. Med.* 4(2): 219–225. https://doi.org/10.4314/ajtcam.v4i2.31211.

Mottaz, J.H., P.J. McKeever, J.L. Neveaux and A.S. Zelickson. 1988. Transdermal delivery of salicylic acid in the treatment of viral papillomas. *Int. J. Dermatol.* 27(8): 596–600.

Mujumdar, A.M., D.G. Naik, R.J. Waghole, D.K. Kulkarni and M.S. Kumbhojkar. 2000. Pharmacological studies on *Sterculia foetida* leaves. *Pharm. Biol.* 38(1): 13–17.

Mukherjee, B., G. Roy and S. Ghosh. 2009. Development of Denticap, a matrix based sustained release formulation for treatment of toothache, dental infection and other gum problem. *Current Drug Deliv.* 6(2): 199–207.

Munday, D.L. and P.J. Cox. 2000. Compressed xanthan and karaya gum matrices: Hydration, erosion and drug release mechanisms. *Intern. J. Pharmaceut.* 203: 179–192.

Naik, D.G., A.M. Mujumdar, R.J. Waghole, A.V. Misar, S.W. Bligh, A. Bashall and J. Crowder. 2004. Taraxer-14-en-3beta-ol, an anti-inflammatory compound from *Sterculia foetida* L. *Planta Med.* 70(1): 68–69.

Namsa, N.D., H. Tag, M. Mandal, P. Kalita and A.K. Das. 2009. An ethnobotanical study of traditional anti-inflammatory plants used by the Lohit community of Arunachal Pradesh, India. *J. Ethnopharmacol.* 125(2): 234–245.

Nandagoapalan, V., A. Doss and C. Marimuthu. 2016. Ethnobotanical studies on useful plants of Pachamalai hills of Tiruchirappalli district of Tamilnadu. *J. Adv. Sci. Res.* 7(1): 14–19.

Omkar, K., S. Suthari, S. Alluri, A. Ragan and V.S. Raju. 2012. Diversity of NTFPs and their utilization in Adilabad District of Andhra Pradesh India. *J. Plant Stud.* 1(1): 33–46. https://doi.org/10.5539/jps.v1n1p33

Onikanni, A.S., B. Lawal, A.O. Olusola, J.O. Olugbodi, S. Sani, B.O. Ajiboye, O.B. Ilesanmi, M. Alqarni, G. Mostafa-Hedeab, A.J. Obaidullah, G.E. Batiha and A.T.H. Wu. 2021. *Sterculia tragacantha* Lindl leaf extract ameliorates STZ-induced diabetes, oxidative stress, inflammation and neuronal impairment. *J. Inflamm. Res.* 14: 6749–6764.

Ortinau, L.C., K.J. Nickelson, K.L. Stromsdorfer, C.Y. Naik, R.T. Pickering, R.A. Haynes, K.L. Fritsche and J.W. Perfield II. 2013. Sterculic oil, a natural inhibitor of SCD1, improves the metabolic state of obese OLETF rats. *Obesity.* 21(2): 344–352.

Ouédraogo, M., K. Konaté, P. Zerbo, N. Barro and L.L. Sawadogo. 2013. Phytochemical analysis and in vitro antifungal profile of bioactive fractions from *Sterculia setigera* (Sterculiaceae). *Current Res. J. Biol. Sci.* 2: 75–80.

Padil, V.V. and M. Černík. 2013. Green synthesis of copper oxide nanoparticles using gum karaya as a biotemplate and their antibacterial application. *Int. J. Nanomedicine.* 8(1): 889–898.

Park, C.R. and D.L. Munday. 2004. Evaluation of selected polysaccharide excipients in buccal adhesive tablets for sustained release of nicotine. *Drug Dev. Ind. Pharm.* 30(6): 609–617.

Partyka, A. 1963. Salad dressing. US Patient No. 3,093,485

Patel, R. and A. Varshney. 2017. Etheno-medicinal plant of Saputara Hill, Dang District, Gujarat. *Intern. J. Scientific Res. Management* 5(9): 6965–6971.

Patra, N., L. Martinová, M. Stuchlik and M. Černík. 2015. Structure-property relationships in *Sterculia urens*/polyvinyl alcohol electrospun composite nanofbres. *Carbohydr. Polym*, 120: 69–73. https://doi.org/10.1016/j.carbpol.2014.12.002

Pattanaik, C., C. Sudhakar Reddy, N.K. Dhal and R. Das. 2006. Some phytotherapeutic claims by tribals of Rayagada District, Orissa, India. *Ethnobotanical Leaflets.* 10: 189–197.

Paul, T. and M.M. Prajapati. 2014. Ethno-therapeutic remedies for bone fracture, Dang Dt, Gujarat. India. *IOSR J. Pharm. Biol. Sci.* 9(2): 45–50.

Postulkova, H., I. Chamradova and D. Pavlinak et al. 2017. Study of effects and conditions on the solubility of natural polysaccharide gum karaya. *Food Hydrocoll.* 67: 148–156. https://doi.org/10.1016/j. foodhyd.2017.01.011

Prajapati, V.D., G.K. Jani, N.G. Moradiya and N.P. Randeria, N.P. 2013. Pharmaceutical applications of various natural gums, mucilages and their modifed forms. *Carbohydr. Polym.* 92(2): 1685–1699.

Prasad, N., N. Thombare, S.C. Sharma and S. Kumar. 2022. Recent development in the medical and industrial applications of gum karaya: A review. *Polym. Bull.* 80: 3425–3447. https://doi.org/10.1007/s00289-022-04227-w

Preetha, B.K. and B. Vishalakshi. 2020. Microwave assisted synthesis of karaya gum-based montmorillonite nanocomposite: characterisation, swelling and dye adsorption studies. *Int. J. Biol. Macromol.* 154: 739–750. https://doi.org/10.1016/j.ijbiomac.2020.03.107

Quattrocchi, U. 2012. CRC world dictionary of medicinal and poisonous plants: Common names, scientific names, eponyms, synonyms, and etymology, [ed. by Quattrocchi, U.]. London, UK: CRC Press Inc., p. 3960.

Rabbi, F., A. Zada, A. Adhikari, A. Jabeen, A. Nisar and I. Ullah. 2017. *Sterculia diversifolia* bears potent anti-cancer and immunomodulatory activities. *Bangladesh J. Pharmacol.* 12: 51–55.

Rabbi, F., A. Zada and A. Nisar. 2020. Larvicidal, leishmanicidal, insecticidal and anthelmintic effects of *Sterculia diversifolia* stem bark and leaf. *Bangladesh J. Pharmacol.* 15(1): 32–38.

Raghavedra Rao, N.G., A. Yadav and U. Kulkarni. 2010. Formulation and evaluation of zero order release glipizide bilayer matrix tablets using natural and synthetic polymers. *Intern. J. Curr. Pharmaceut. Res.* 2(1): 34–42.

Raj, V., J.H. Lee, J.J. Shim and J. Lee. 2021. Recent findings and future directions of grafted gum karaya polysaccharides and their various applications: A review. *Carbohydr Polym.* https://doi.org/10.1016/j.carbpol.2021.117687

Raja, T.A.R., R.V.R. Reddy and K.U.M. Rao. 2014. Evaluation of anticonvulsant effect of *Sterculia foetida* (Pinari) in pentylenetetrazole (PTZ) and mes induced convulsions in albino rats. *World J. Pharma. Pharmaceut. Sci.* 3(3): 1898–1907.

Rani, P.U. and P. Rajasekharreddy. 2009. Toxic and antifeedant activities of *Sterculia foetida* (L.) seed crude extract against *Spodoptera litura* (F.) and *Achaea janata* (L.). *J. Biopestic.* 2(2): 161–164.

Rani, P.U. and P. Rajasekharreddy. 2010. Insecticidal activity of (2*n*-octyl*cyclo*prop-1-enyl)-octanoic acid (I) against three Coleopteran stored product insects from *Sterculia foetida* (L.). *J. Pest. Sci.* 83: 273–279.

Ranta, F., D.S. Nawawi, E.S. Pribadi and W. Syafii. 2012. Antifungal activity of Faloak (*Sterculia comosa* Wallich) extractives. *J. Trop. Timber Sci. Technol.* 10(1): 60–65.

Rastogi, R.P. and B.N. Mehrotra. 1993. *Compendium of Indian medicinal plants,* Vol. 3, New Delhi: Publication and Information Directorate, pp. 609–610.

Ravi Prasad Rao, B. and S. Sunitha. 2011. Medicinal plant resources of Rudrakod Sacred Grove in Nallamalais, Andhra Pradesh, India. *J. Biodiversity* 2(2): 75–89.

Riya, K., B.K. Jain and R.S. Patel. 2020. Ethnobotanical study of medicinal plants used to treat human diseases in Bhambhori and Itawa forest areas of Dahod, Gujarat, India. *Plant Archiv.* 20: 222–228.

Rollando, R. and R. Alfanaar. 2017. Cytotoxic effect of 2,3-dihydro-6-hydroxy-2- methylenenaphtho[1,2-b] furan-4,5-dione compound from the bark of faloak (*Sterculia quadrifida* R.Br.) in breast cancer cells T47D. *Pharmaciana* 7(2): 289–294.

Rollando, R. and E. Monica. 2018. Determination of total phenolic content and antioxidant activity test of water fraction methanol extract of Faloak stem bark (*Sterculia quadrifida* R.Br). *Scientia : J Pharmacy Health* 8(1): 29–36.

Rollando, R. and K.R. Prilianti. 2017. Ethyl acetate fraction of faloak (*Sterculia quadrifida* R.Br.) bark induces apoptosis and cell cycle on T47D breast cancer cells. *J Pharmaceut. Sci. Commun.* 14(1): 1–14.

Rollando, R. and S. Siswadi. 2016. Investigation of the potential cytotoxic activity of the stem bark fraction of faloak plants (*Sterculia quadrifida* R.Br). *J. Pharmaceut. Sci. Clin. Pharma.* 13(1): 27–32.

Rollando, R., R. Sitepu and E. Monica. 2018. Cytotoxic activity of 2- iminoethyl 2-(2-(1-hydroxypentan-2-yl) phenyl) acetate from *Sterculia quadrifida* R.Br. ethyl acetate fraction. *J. Global Pharma. Technol.* 10(6): 204–212.

Rollando, R., Y. Yuniati, R.A. Sanggen, E. Monica and H. Setiawan. 2019. Effects of solvent composition on profile of total extracts, phenolic compounds, and antifungal activity of *Sterculia quadrifida* R.Br bark. *J. Global Pharma Technol.* 11(1): 342–351.

Rupesh, M. and D. Nitin. 2012. Studies on the medicinal uses of wild trees of Nagpur District. *Intern. J. Life Sci. Pharma Res.* 2(1): L21–L24.

Saefudin, S., S. Marusin and C. Chairul. 2013. Antioxidant activity on six species of Sterculiaceae plants. *Forest Sci. Res. J.* 31(2): 103–109.

Sanderson, G.R. 1996. Gums and their use in food systems. *Food Technol.* 50: 81–84.

Sanjay, V.S. 2020. Mono-herbal treatments for wound healing. *Intern J Curr Res.* 12(1): 9823–9827

Saragih, G.S. and S. Siswadi. 2019. Antioxidant activity of plant parts extracts from *Sterculia quadrifida* R. Br. *Asian J. Pharmaceut. Clin. Res.* 12(7): 143–148.

Sarathchandiran, I. and P. Suresh Kumar. 2013. Characterization and standardization of gum karaya. *Int. J. Biopharm.* 4: 180–189.

Sawayanagi, Y., N. Nambu and T. Nagai. 1983. Dissolution properties and bioavailability of phenytoin from ground mixtures with chitin or chitosan. *Chem. Pharm. Bull. (Tokyo)* 31: 2064–2068. https://doi.org/10.1248/cpb.31.2064

Sawmliana, M. 2013. The book of Mizoram plants. 2nd ed. Chanmari: Lois Bet.

Senapati, M.K., A. Srinatha, J.K. Pandit. 2006. In vitro release characteristics of matrix tablets: Study of karaya gum and guar gum as release modulators. *Indian J. Pharmaceut. Sci.* 68(6): 824–827.

Shah, R., M. Islam, F. Rabbi, S. Nasrin Akter, A. Akter, H. Akter, I. Ahmed, E. Islam and Md. Rahmatullah. 2017. Phytotherapeutic practices of a folk medicinal practitioner in Dinajpur district, Bangladesh. *J. Appl. Pharmaceut. Sci.* 7(5): 161–165.

Shazia Hussain, S., M. Janarthan and S. Anusha. 2014. Preclinical evaluation of anti-diabetic and antihyperlipidemic activity of methanolic extract of *Sterculia foetida* leaves by using Wistar albino rats. *Indian J. Res. Pharma Biotechnol.* 2(6): 1430–1438.

Shukla, A., K. Desai and N. Modi. 2020. *In vitro* antioxidant and antimicrobial potential of *Sterculia urens* Roxb. root extract and its bioactive phytoconstituents evaluation. *Futur. J. Pharm. Sci.* 6: 45. https://doi.org/10.1186/s43094-020-00063-9

Silva, F.V., I.S. Oliveira, K.A. Figueiredo et al. 2014. Anti-inflammatory and antinociceptive effects of *Sterculia striata* A. St.-Hil. & Naudin (Malvaceae) in rodents. *J. Med. Food.* 17(6): 694–700.

Silva, S.C.C.C., E.M. de Araujo Braz, F.A. de Amorim Carvalho, C.A.R. de Sousa Brito, L.M. Brito, H.M. Barreto, E.C. da Silva Filho and D.A. da Silva. 2020. Antibacterial and cytotoxic properties from esterified *Sterculia* gum. *Intern. J. Biol. Macromolecules* 164: 606–615.

Singh, B. and A. Kaur. 2011. Comparison of agar and gum karaya as gelling agent for in vitro regeneration of rough lemon (*Citrus jambhiri* Lush.) plantlets from nodal explants. *J. Crop Sci. Biotechnol.* 14: 297–303. https://doi.org/10.1007/s12892-011-0037-z

Singh, B. and L. Pal. 2008. Development of *Sterculia* gum based wound dressings for use in drug delivery. *European Polymer J.* 44(10): 3222–3230.

Singh, B. and N. Sharma. 2008. Development of novel hydrogels by functionalization of *Sterculia* gum for use in anti-ulcer drug delivery. *Carbohydr. Polym.* 74: 489–497. https://doi.org/10.1016/j.carbpol.2008.04.003

Singh, B. and N. Sharma. 2009. Mechanistic implication for cross-linking in sterculia-based hydrogels and their use in GIT drug delivery. *Biomacromol* 10: 2515–2532. https://doi.org/10.1021/bm900 4645

Singh, B. and N. Sharma. 2011. Design of *Sterculia* gum based double potential antidiarrheal drug delivery system. *Colloids and Surfaces B: Bio Interfaces* 82(2): 325–332.

Singh, B., V. Sharma and D. Chauhan. 2010. Gastroretentive floating *Sterculia*–alginate beads for use in antiulcer drug delivery. *Chem. Eng. Res. Design* 88(8): 997–1012.

Singh, B., V. Sharma and A.R. Kumar. 2021. Designing moringa gum-*Sterculia* gum-polyacrylamide hydrogel wound dressings for drug delivery applications. *Carbohydrate Polymer Technologies and Applications* 2: 100062.

Singh, B., V. Sharma and L. Pal. 2011. Formation of sterculia polysaccharide networks by gamma rays induced graft copolymerization for biomedical applications. *Carbohydr. Polym.* 86: 1371–1380. https://doi.org/10.1016/j.carbpol.2011.06.041

Singh, B. and B. Singh. 2018. Modification of sterculia gum polysaccharide via network formation by radiation induced cross linking polymerization for biomedical applications. *Int. J. Biol. Macromol.* 116: 91–99. https://doi.org/10.1016/j.ijbiomac.2018.05.032

Singh, B. and M. Vashishtha. 2008. Development of novel hydrogels by modification of sterculia gum through radiation cross-linking polymerization for use in drug delivery. *Nucl. Instrum. Methods Phys. Res. Sect B Beam Interact with Mater. Atoms.* 266: 2009–2020. https://doi.org/10.1016/j.nimb.2008.03.086

Singh, C.B. 2017. Rare and threatened medicinal plants of Santhal Pargana, Jharkhand. *J. Indian Bot. Soc.* 96: 94–99.

Singh, S., J. Singh, M.S. Muthu, J. Balasubramaniam and B. Mishra. 2007. Gastroretentive drug delivery system of metoclopramide hydrochloride: formulation and in vitro evaluation. *Curr. Drug Deliv.* 4(4): 269–275.

Sirisha, V.N.L., M.C. Eswariah and A. Sambasiva Rao. 2017. Formulation and evaluation of karaya gum microspheres of cromolyn sodium: Treatment of ulcerative colitis. *Int. Res. J. Pharm.* 8(10): 182–188.

Sousa, J.A., I.S. Oliveira, F.V. Silva, D.A. Costa, M.H. Chaves, F.A. Oliveira, P.H. Nunes and R.C. Oliveira. 2012. Gastroprotective activity of *Sterculia striata* A. St. Hil. & Naudin (Malvaceae) in rodents. *Z. Naturforsch. C. J. Biosci.* 67(3–4): 163–171.

Sparberg, M., J.V. Prohaska and J.B. Kirsner. 1966. Solid state karaya gum ring for use in disposable and permanent ileostormy application. *Amer. J. Surg.* 111: 610–611.

Sreekanth Reddy, O., M. Subha, T. Jithendra, C. Madhavi and K. Chowdoji Rao. 2019. Fabrication of Gelatin/Karaya gum blend microspheres for the controlled release of Distigmine bromide. *J. Drug Delivery and Therapeutics* 9(3–s): 1–1.

Srivastava, S. and D.P. Ray. 2015. Natural gum resources in India and their commercial importance. *Intern. J. Bioresource Sci.* 2(2): 151–155.

Steinhardt, A. and F.A. Goldwater. 1962. Gelatin adhesive pharmaceutical preparations. US Patent, 3,029,187

Sujja-areevatha, J., D.L. Mundaya, P.J. Cox and K.A. Khan. 1998. Relationship between swelling, erosion and drug release in hydrophillic natural gum mini-matrix formulations. *European J. Pharmaceut. Sci.* 6: 207–217.

Sukumaran, S. and A.D.S. Raj. 2010. Medicinal plants of sacred groves of Kanyakumari district, Southern Western Ghats. *Indian J. Trad. Knowl.* 9(2): 294–299.

Suryanarayana Swamy, N. and T.V.V. Seetharami Reddi. 2016. Ethnomedicine for rheumatoid arthritis by the tribes of Adilabad district, Andhra Pradesh. *J. Trad. Folk Pract.* 2(3): 68–77.

Tamilselvi, Nivitha, Priya and Monika. 2019. A study to assess the knowledge of karaya powder application on diabetic wound healing among diabetic mellitus patients. *Int. J. Res. Pharmaceut. Sci.* 10(4): 2701–2704.

Tania, K.N., M.T. Islam, A. Mahmood, M. Ibrahim, M.M.U. Chowdhury and M.R. Kuddus. 2013. Pharmacological and phytochemical screenings of ethanol extract of *Sterculia villosa* Roxb. *J. Biomed. Pharmacol.* 2: 9–14.

Taupin, P.J. and D.M. Anderson. 1982. Sub-chronic toxicity study in rats fed gum karaya. *Food Chem. Toxicol.* 20(5): 513–517.

Tor-Anyiin, M.T.A., U. Akpuaka and O.A.H. Oluma. 2011 Phytochemical and antimicrobial studies on stem bark extract of *Sterculia setigera*, Del. *African J. Biotechnol.* 10(53): 11011–11015.

Udegbunam, R.I., I.U. Asuzu, R.O.C. Kene and S.O. Udegbunam. 2013. Evaluation of local anaesthetic efficacy of the crude extract of *Sterculia tragacantha* using West African Dwarf Goats. *Sokoto J. Vet. Sci.* 11: 13–21.

Udegbunam, R.I., I.U. Asuzu, R.O.C. Kene, S.O. Udegbunam and C.O. Nwaehujor. 2011. Anti-nociceptive, anti-inflammatory and anti-oxidant effects of the methanol leaf extract of *Sterculia tragacantha* Lindl. *J. Pharmacol. Toxicol.* 6: 516–524.

Udegbunam, R.I., I.U. Asuzu, R.O.C. Kene, S.O. Udegbunam and C.O. Nwaehujor. 2012. Local anesthetic and tissue effects of the leaf extract and fractions of *Sterculia tragacantha* Lindl. *J. Pharmacol. Toxicol.* 7(4): 192–198.

Upadhyay, R.K. 2017. Nutritional, therapeutic, and pharmaceutical potential of plant gums: A review. *Int. J. Green Pharm.* 11: S30–S48.

Venkata Ratnam, K. and R.R. Venkata Raju. 2005. Folk medicine used for common women ailments by Adivasis in Eastern Ghats of Andhra Pradesh. *Indian J. Trad. Knowl.* 4(3): 267–270.

Verbeken, D., S. Dierckx and K. Dewettinck. 2003. Exudate gums: occurrence, production, and applications. *Appl. Microbiol. Biotechnol.* 63: 10–21.

Vilas, V.B. 2021. Few Ethno-Botanical Medicinal Plants from Toranmal Plateau, Maharashtra, India. *J. Scientific Res.* 65(7): 99–102.

Vinod, V.T.P., S. Chandra and C. Miroslav. 2015b. Dodecenylsuccinic anhydride derivatives of gum karaya (*Sterculia urens*): Preparation, characterization, and their antibacterial properties. *J. Agric. Food Chem.* 63(14): 3757–3765.

Vinod, V.T.P., H.A.N. Nhung, S. Alena and Č. Miroslav. 2015a. Fabrication, characterization, and antibacterial properties of electro-spun membrane composed of gum karaya, polyvinyl alcohol, and silver nanoparticles. *J. Nanomaterials.* ID 750726, 10. https://doi.org/10.1155/2015/750726

Vinod, V.T.P., R.B. Sashidhar, N. Sivaprasad et al. 2011. Bioremediation of mercury (II) from aqueous solution by gum karaya (*Sterculia urens*): A natural hydrocolloid. *Desalination* 272: 270–277. https://doi.org/10.1016/j.desal.2011.01.027

Vinod, V.T.P., S. Wacławek, C. Senan et al. 2017. Gum karaya (*Sterculia urens*) stabilized zero-valent iron nanoparticles: Characterization and applications for the removal of chromium and volatile organic pollutants from water. *RSC Adv.* 7: 13997–14009. https://doi.org/10.1039/c7ra00464h.

Vital, P.G., R.N. Jr. Velasco, J.M. Demigillo and W.L. Rivera. 2010. Antimicrobial activity, cytotoxicity and phytochemical screening of *Ficus septica* Burm. and *Sterculia foetida* L. leaf extracts. *J. Med. Plants Res.* 4(1): 58–63.

Wang, R.F., X.W. Wu and D. Geng. 2013. Two cerebrosides isolated from the seeds of *Sterculia lychnophora* and their neuroprotective effect. *Molecules.* 18(1): 1181–1187. https://doi.org/10.3390/molecules18011181.

Waters, D., D. Freedman, J. Lesperance and P. Theroux. 1987. Design features of a controlled clinical trial to assess the effect of a calcium entry blocker upon the progression of coronary artery disease. *Control Clin. Trials.* 8: 216–242.

Weeks, J.M. and N.J. Black. 1992. Notes on the ethnopharmacology of the Lenca Indians of western Honduras and eastern El Salvador. *Mexicon.* 14(4): 71–74.

Weiping, W. and A. Branwell. 2000. Tragacanth and Karaya. In: Philips, G.O. and P.A. Williams (eds.) Hand book of hydrocolloids. Amsterdam: Elsevier, pp 231–246

Whistler, R.L. 1993. Exudate gums. Industrial gums: Polysaccharides and their derivatives. 3rd ed. Amsterdam: Elsevier Inc., pp. 309–339.

Wilson, E., M. Chacha and J. Omolo. 2014. Antimicrobial and cytotoxic activities of extracts from *Sterculia quinqueloba* (Garcke) K. Schum and *Canthium crassum* Hiern. *J. Pharmacogn. Phytochem.* 3(3): 125–129.

Wilson, M. and W. Harvey. 1989. Prevention of bacterial adhesion to denture acrylic. *J. Dent.* 17(4): 166–170.

Wu, Y., S.W. Cui, J. Tang, Q. Wang and X. Gu. 2008. Extraction, characterisation and anti-inflammatory bioactivity of polysaccharides from boat-fruited *Sterculia* seeds. In Williams, P.A. and G.O. Phillips (eds.) *Gums and stabilisers for the food industry.* Cambridge: Royal Society of Chemistry, pp. 367–378.

Yang, Y., B.I. Park, E.-H. Hwang and Y.O. You. 2016. Composition analysis and inhibitory effect of *Sterculia lychnophora* against biofilm formation by *Stereptococcus nutans. Evid. Based Complent. Alternat. Med.* 2016: 8163150.

7 Propagation of *Sterculia urens* and Other Species of *Sterculia*

K. Venkata Ratnam,
Lepakshi Md. Bhakshu, and T. Pullaiah

7.1 INTRODUCTION

Gums and resins occupy a prime place among non-wood forest produce (NWFP/NTFP) and are known to mankind since time immemorial. Gum trees are economically important and found in tropical moist and dry deciduous forests and produce a significant quantity of gum, which are widely used for industrial, food, and medicinal purposes in India. Tree gum exudates of the natural acid polysaccharide have attracted significant attention among researchers because of their immense potential application in the food industry, biomedicine, and material science. Among the different tree gum polysaccharides, gum karaya (GK) is an important partially acetylated natural polysaccharide.

The relatively slow growth of tree and the tapping of gum at various stages of tree growth hinder the overall performance of the species. Exploitation of gum with unscientific tapping techniques is another cause of its depletion. Individuals outside the protected areas are comparatively more prone to become threatened due to a lack of protection to the species. In the absence of cultivation of these trees in regular plantation, there is grave concern about the loss of wild germplasm of *S. urens*. Hence, there is an urgent need for more propagation studies on the gum karaya tree.

7.2 PHENOLOGY

Trees shed leaves in November to December and remain leafless till May, while new leaves appear from April to June. Flowers are polygamous. The flowers in crowded, erect, pyramidal panicles appear from December to March before leaf flushing. The tree produces a large number of male and a smaller number of bisexual (functionally female) flowers. The species is self-incompatible, and fruits set by open pollination are often poor, probably because of pollination limitations. Fruit of 4–5 sessile radiating follicles are covered outside with stiff, stinging bristles, orange to bright red when ripe. Seeds 3 to 6 in each follicle, 6 mm long, oblong, dark brown, and shining (Figure 7.1). The follicles ripen about April to May and split in hot and dry weather. Seed dispersal is by birds (Sukhadiya et al., 2019).

 DOI: 10.1201/9781003438960-7

FIGURE 7.1 Seeds of *Sterculia urens*.

Source: www.indiamart.com/proddetail/sterculia-urens-seed-22752815612.html.

Before seed collection, the ground is kept clean and the follicles are allowed to burst on the tree, and then seeds are collected from the ground. Otherwise, the follicle bunch is knocked off from the tree and dried in the sun so that the seeds burst out. The handling of the follicles becomes difficult because of the stinging bristles. Seed can be stored easily for one year.

Fresh fruits are collected from the trees. Follicles should be collected 1 1/2 to 2 months after anthesis in the last week of April to the first week of May, when the fruit colour is yellowish green and seed colour is mostly black or grey. The collection method is to spread a tarpaulin under the tree and collect the fruits by lopping the branches or plucking. After collection, seeds should be extracted from the fruits; further, white and brown seeds should be discarded just after collection, as some may turn grey under ambient condition and therefore become indistinguishable from high-quality grey seeds. After extraction, black and grey seeds should be dried under shade to reduce moisture content and attain maximum germinability (Sukhadiya et al., 2019).

Seed Biology:

No. of seeds per kg	*Germination percentage*	*Plant percent*	*Time taken for germination in days*	*No. of seedlings per kg of seed*
5,500 to 6,360	32 to 77	18 to 90	10 to 15	1,150 to 1,500

7.3 SEED GERMINATION

Seeds are orthodox type and tolerate desiccation to 4–5% moisture content. Seeds will remain viable at least for two years, if stored with 4–8% moisture content at ambient temperature. Seed viability can be maintained more than three years at 15°C or below. In moisture content 10% and above, seed germination is decreased if stored above 15°C. Germination can be improved by removing two layers of seed coat (grey and white). Seeds are filled in cloth bag and rubbed to remove the two layers, and then the seeds are washed and dried in the shade (Sukhadiya et al., 2019).

The effect of storage temperature and dormancy-breaking treatments on seed germination, moisture content, and seed vigour in gum karaya (*Sterculia urens*) was investigated by Subhashini Devi et al. (2012). Seeds of *Sterculia* species have a hard seed coat that is impermeable to water, and consequently, the dispersal unit has physical dormancy. In *S. urens*, dormancy might be due to a hard seed coat that is impermeable to water. Acid treatment of 4 min may allow water, oxygen, or both to enter the seeds of *S. urens*, thus permitting the embryo to overcome the mechanical restriction of the surrounding tissues. When these types of treatments break dormancy, it is considered as physical dormancy, since the removal of seed coat allows the penetration of water for germination. The response of *S. urens* to H_2SO_4 treatment indicates physical dormancy (Subhashini Devi et al., 2012).

Pre-treatment with GA_3 significantly enhanced germination in *S. urens*. GA_3 at 150 ppm yielded the highest germination percentage when compared with BAP and IAA. GA_3 has been found to be effective in increasing germination in several species and to break dormancy in dormant seeds. Embryo dormancy is thought to be due to the presence of inhibitors, especially ABA, as well as the absence of growth promoters, such as GA_3. The loss of embryo dormancy is often associated with a sharp drop in the ratio of ABA to GA_3. These findings suggest that exogenous application of GA_3 releases embryo dormancy and promotes germination and counteracts inhibitory effects of ABA directly or indirectly in *S. urens*. Enhancement of germination with hormones clearly indicated physiological dormancy in *S. urens* (Subhashini Devi et al., 2012). The results of temperature and light treatment showed that germinating seeds at 16 hours daylight and 8 hours dark at room temperature (30 + 28°C) after 24 hours soaking was best for getting a better germination percentage. These findings also indicate that light has a role in increasing seed germination in *S. urens*.

Sukhadiya et al. (2019) gave the details of seed germination. Germination is epigeal. The species can be raised by direct sowing in lines 30 cm wide and 1.8 m apart. Seeds may be sown in patches at a spacing of 1.8 m × 1.8 m, with three seeds in each patch. Sowing is done soon after collection at the onset of rains. Germination takes place in 10–15 days. Preferred planting material is nursery-raised plants with balls of earth or bagged plants. The seedlings should be planted during the onset of rains in June to July. The species is intolerant to even short-term waterlogging. The period of establishment in the field can be reduced by applying chemical fertilizers, for example, 25 g of a mixture of ammonium sulphate and superphosphate per plant applied in two dosages with three weeks' interval. The seedlings and saplings may be damaged by fire, so clear weeding around the plants is desirable. Seedlings should be protected from damage by cattle, sambhar, chital, and rodents.

The right choice of the growing medium is one of the most important factors influencing germination and growth. Mohammad et al. (2021) screened and identified suitable media for germination of *S. urens*. Sand, soil, and mixture of 'sand + soil + FYM' in equal proportion were screened following completely randomized design for its effect on final germination percentage and associated parameters. Based on the estimates of final germination percentage (FGP) and associated parameters, namely, mean germination time (MGT), coefficient of velocity of germination (CVG), germination rate index (GRI), germination index (GI), peak value of mean germination (PVMG), time spread of germination (TSG), and energy period (EP), it is concluded that sand is the appropriate medium for germinating seed of *S. urens* in mass scale. It gives the highest germination percentage with faster rate, resulting in saving of time and nursery resources. Correlation coefficient analysis among studied germination parameters was also worked out and found that relationship is in agreement with the definitions of respective parameters.

For germination of seeds, overnight soaking in ordinary water is sufficient. Treated seeds are sown in poly bags kept on hard surface in May without any shading. After germination, watering is done once a day, and copper fungicide is applied to ward off damping of seedlings. Then, 1-month-old seedlings are planted out.

7.4 PROPAGATION OF *STERCULIA URENS*

Sterculia urens is a plant of low elevation in the arid to moist tropics, where it is usually found at elevations from 300 to 750 m. It grows best in areas where annual daytime temperatures are within the range 30–42°C, though it can tolerate 7–47°C. It can be killed by temperatures of –2°C. It prefers a mean annual rainfall within the range 750–1,200 mm but tolerates 200–2,000 mm, usually growing in areas with a distinct dry season. It requires a sunny position, though seedlings are shade-tolerant. It grows best in a light, well-drained soil. It grows in sandstone, quartzite (not on laterite) rock. It prefers ferruginous and rocky soils but avoids loam or heavier soils. Very tolerant of poor, rocky soils. Established plants are very drought-tolerant. Performs best on free-draining loam and sand soils of a moderately acid to slightly alkaline nature, generally with a pH of 5.5 to 7.5, and on sites with full to partial sun exposure; prefers a pH in the range 6–7 but tolerates 5.5–7.8. The tree has poor tolerance to slow-draining or saturated soils and shade conditions.

New plants are usually started from seed germination. Seed germination takes 10–15 days and gives nearly 100% seedlings that reach 15–20 cm in height in three months (*www.worldagroforestry.org/*). A physical dormancy caused by the hard seed coat of the mature seeds of many species in this genus can be overcome by scarifying the seed. This is carried out by cutting away or abrasing some of the seed coat to allow the ingress of water, though great care must be taken not to damage the embryo (*http://en.sl.life.ku.dk/Publikationer/Udgivelser/PopulaerPublikationer.aspx?katid={D28373CC-6EF3-4EF8-B097-6D83FABF209E}&serieid={9F1C3DB1-6E7B-4CF1-AF53-F480B0CB40EF}&sort=title*). The aril surrounding the seed should also be removed – this is easiest when it has been softened through soaking in water. The seeds germinate optimally at temperatures between 20 and 30°C. They can be sown in a nursery seedbed or in containers. A germination rate of about 95%, occurring within about two weeks, can be expected if the seed has been properly treated.

Rayudu (2003) described methods of nursery raising of seedlings of *Sterculia urens*. The seeds were sown in the nursery. In the primary beds, large-scale plants of *S. urens* were raised. Around 4 kg of seeds were sown on 22 primary beds. 7,500 seedlings were collected from the nursery and planted in forestlands near Surendrapalem and Solab Village. Seedlings from a few days old to 3 weeks old were transplanted into polythene bags. It was found that few-days-old seedlings transplanted into polythene bags resulted in 80% of survival.

Regular watering, providing thatched shade helped for the above 80% survival rate both in primary beds and secondary beds. After the fourth day of sowing seed, sprouting was observed. In the species, radicle emergence was noticed within 5 days' period, and plumule emergence was noticed within 6–7 days. Species germination took 8 days. During the radicle emergence, a small bulb-like structure is formed on the radicle. These bulbs become a tuber when the species growth increases. The tuber is developed just below the collar region. It has been observed that *Sterculia urens* seedlings did not tolerate waterlogged conditions. However, 2 to 3-month-old seedlings can withstand drought conditions.

During waterlogging conditions in the species, the tuber is damaged. Rayudu et al. (2003) reported that in their experiment, ten seedlings were transplanted in earmarked waterlogged area, and it was found that five seedlings were affected due to three-day waterlogging conditions. These seedlings lost leaves and finally died. On further observation, it was noticed that the tubers were decayed. Hence, it can be safely assumed that the damage of the tubers is due to disconnection between the root system and shoot system. In the nursery, seed density experiment was conducted. For this experiment, *Sterculia urens* seed 150, 200, 250 g were sown on three separate beds. The size of these beds was 1×1 m^2. It was found that 150 g seeds' bed density was sufficient for seedling growth.

The trees can be tapped for their gum about five times during their lifetime. The trees are generally worked under some form of coppice system on a rotation of 30–40 years.

7.5 VEGETATIVE PROPAGATION

Terminal and middle cuttings of *Strculia urens* taken from selected trees of 70–80 cm girth in a reserved forest at Godavari, Andhra Pradesh, India, in April 1992 (at the leaf shedding stage) were treated with IBA at 4,000 or IAA at 5,000 ppm. Similar rooting of cuttings was observed after 28 days in both growth regulator treatments, but there was no rooting in the control without either.

Rayudu (2003) described methods of vegetative propagation. Under the experiment vegetative propagation, *Sterculia urens* stumps were collected from high-yielding gum tree. The stumps were planted in the nursery. Fifty stumps were 12 cm long, 20–25 mm in girth. Fifty stumps were 12 cm long and above 30 mm in girth. Twenty stumps were 1 1/2 ft in length and 60 mm in girth. Around 30–40% survival was recorded from big-size 20 sturdy wood stumps. However, no survival was recorded from the 12 cm long and 20–25 mm girth stumps. It was observed that only 5 stumps survived from 12 cm long and above 30 mm size stumps. These stumps were rejuvenated after 20 days. No hormone treatment given for stumps.

7.6 PROPAGATION OF OTHER *STERCULIA* SPECIES

7.6.1 *Sterculia alata* (Roxb.) R.Br.

S. alata is commonly known as Buddha's coconut tree. The effect of different chemical, physical, and mechanical methods of breaking the dormancy of seeds of Buddha's coconut was evaluated. In physical method, seeds were treated with hot water for 24 and 36 hours. In chemical methods, seeds were treated with 1% and 2% H_2SO_4 for 10 min, KNO_3 1% and 2% for 10 min, 1% and 2% thiourea for 10 min, and GA_3 at three different concentrations, namely, 750, 500, and 250 ppm. In mechanical method, the seeds were scarified for 3 min. Seeds treated with hot water, H_2SO_4, and mechanical scarification demonstrated significantly higher germination than other treatments. Seeds treated with chemical methods showed various degrees of germination (Khan, 2016).

7.6.2 *Sterculia alexandri* Harv.

Sterculia alexandri is a medium-size tree endemic to Eastern Cape, west of Port Elizabeth of South Africa. The seeds sown in well-drained sandy soil showed good germination. Germination occurs within three weeks' time. After germination, the root forms a small bulb that later will become tuber. The seedlings grow very slowly (Palmer and Pitman, 1972). The plant also is propagated by vegetative means through stem cuttings in spring and summer seasons. Stem cuttings of 150–300 mm in length are ideal for vegetative propagation. The cuttings grow in soil fed with compost or organic fertilizer (Palmer and Pitman, 1972).

7.6.3 *Sterculia chicha* A. St. Hil.

Tavares et al. (2013) assessed the effect of *Anastrepha bezzii* larvae association (Figure 7.2) on germination of *Sterculia chicha* seeds using four treatments: (1) intact seeds without scarification, (2) intact seeds scarification with sandpaper for 3 min, (3) drilled seeds without scarification, and (4) drilled seeds scarified with sandpaper for 3 min. Among the tested treatments, intact seeds scarification with sandpaper for 3 min showed the highest germination. The intact seeds without scarification and drilled seeds scarified with sandpaper for 3 min groups showed moderate activity, and drilled seeds without scarification showed the lowest germination (Tavares et al., 2013).

7.6.4 *Sterculia foetida* L.

Propagation of *S. foetida* through seeds collected from Indira Gandhi Botanical Garden (Figure 7.3a,b) was studied by Rai (2014). About 100 healthy and mature seeds of *S. foetida* were sown in cement pots (10 seeds for each pot) filled with soil and manure (3:1). Germination started from the 7th day onwards, and maximum

FIGURE 7.2 *Sterculia chicha* seeds damaged by *Anastrepha bezzii*.

Source: Tavares et al. (2013).

FIGURE 7.3 (a) *Sterculia foetida* pods with seeds; (b) seeds separated from pods.

Source: Rai (2014).

FIGURE 7.4 (a) Germination of *Sterculia foetida* seeds (B): Six months old seedlings

germination (70%) was reached on the 21st day. Sapling growth was recorded at two months' interval. Six-month-old plants were transplanted in urban and rural areas of Meerut District (Figure 7.4a,b) (Rai, 2014). Salim Azad et al. (2018) conducted experiments to find out the effect of IBA at different concentrations (0.2. 0.4, 0.6, and 0.8%) on rooting of branch cuttings collected from leafy and non-leafy branches of *S. foetida*. The highest rooting was observed in 0.8% IBA-treated both leafy and non-leafy stem cuttings. Lowest rooting was observed in control groups of both cuttings.

7.6.5 *Sterculia quadrifida* R.Br.

A study was conducted to examine the effect of donor plant or rooting medium on the propagation of *Sterculia quadrifida* through stem cuttings collected from juvenile and mature plants. Stem cuttings collected from seedlings are 3.7–8.5 mm in diameter, while stem cuttings collected from trees had two sizes, namely, big size (15.4–24.4 mm) and small size (5.8–13.8 mm). Three types of growth media were used for propagation: (1) only soil, (2) equal volume of soil and manure, and (3) coco peat. The results demonstrated that stem cuttings with big size, one larger diameter, showed more survival rates than stem cuttings of smaller diameter (Rianawati and Siswadi, 2020). The effects of climatic factors on the survival rate of *S. quadrifida* stem cuttings of different length (9, 18, and 27 cm) propagated in different root media, such as soil, soil with manure, soil with sand, sand and coco peat, were investigated. The results demonstrated that stem cuttings of 27 cm grown in rooting medium of soil and coco peat showed the highest survival rate (Figures 7.5 & 7.6). Low humidity and high temperature cause low survival rate of stem cuttings (Siswadi et al., 2017).

FIGURE 7.5 Formation of buds after one week of transplantation of *S. quadrifida* stem cuttings.

Source: Rianawati and Siswadi (2020).

FIGURE 7.6 Formation of leaves in two weeks of transplantation of *S. quadrifida* stem cuttings.

Source: Rianawati and Siswadi (2020).

7.6.6 *Sterculia appendiculata* K.Schum.

Sterculia appendiculata is a deciduous tree distributed in east tropical Africa – Kenya, Tanzania, Malawi, Mozambique. Seeds collected from ripen fruits are good for sowing. Scarification is needed to overcome physical dormancy due to hard coat. The aril present around the seed should be removed for easy germination of seed. The optimum temperature for seed germination is 20–30°C. The germination time takes two weeks, and germination showed 95%.

7.7 CONCLUSION

The existence of plant species depends on the capacity of its seed vigour, and propagative ability makes species flourish for several generations, even population spreading. The studies on the seed germination by scientists from different parts of the globe elucidate and reveal that *Sterculia* species have full potency of germination. However, certain pest and seed-eating animals make non-availability of seed for germination and slow spreading. Moreover, vegetative methods for the propagation are not enough and need to be investigated, since it is easy and cost-effective. Hence, certain silvicultural practices have to be developed for mass propagation from the genetically rich ideal plants to encourage socio-forestry as well as to conserve endangered *Sterculia* species.

REFERENCES

www.indiamart.com/proddetail/sterculia-urens-seed-22752815612.html

Khan, M.R. 2016. Pre-germinative treatments to overcoming of seed dormancy in *Sterculia alata* (Roxb.) *Clompanus alata* (Roxb.) Kuntze. *Bio Science Research Bulletin* 32(1 & 2): 39–43.

Mohammad, N., A. Dahayat and Y. Pardhi. 2021. Effect of growing media on seed germination attributes in *Sterculia urens* Roxb.—An endangered gum yielding tree of tropical forest. *Indian Forester* 147(7): doi: 10.36808/if/2021/v147i7/153636

Palmer, E., and N. Pitman. 1972. *Trees of southern Africa.* Cape Town: Balkema.

Rai, Y. 2014. Growth and development of rare tree species *Sterculia foetida* Linn. in District Meerut (U.P.) India. *International Journal of Scientific Research* 3: 324–327.

Rayudu, V. 2003. Propagation and scientific collection of gum from *Sterculia* for sustainability. In: *XII world forestry congress.* Quebec City: Food and Agricultural Organization, pp. 0229–B2.

Rianawati, H., and Siswadi. 2020. Effect of donor plants and rooting medium on stem cutting propagation of faloak (*Sterculia quadrifida*). *Tropical Drylands* 4: 31–35.

Salim Azad, Md., Md. Jahidul Alam, Mollick and Md. N.I. Khan. 2018. Rooting of cuttings of the wild Indian almond tree (*Sterculia foetida*) enhanced by the application of indole-3-butyric acid (IBA) under leafy and non-leafy conditions. *Rhizosphere* 5: 8–15.

Siswadi, G.S. Saragih, R. Setyowati and R.Y. Puspiyatun. 2017. Constraints in stem cuttings propagation of Faloak (*Sterculia quadrifida* R.Br.) in high temperature region. *Proceedings of the IUFRO–INAFOR Joint International Conference—POSTER C* 7: 385–390.

Subhashini Devi, P., B. Satyanarayana, A. Arundhati and T. Raghavarao. 2012. Effect of storage temperature and dormancy-breaking treatments on seed germination, moisture content and seed vigor in gum karaya (*Sterculia urens* Roxb.). *Forest Science Tech* 8(1): 1–5. DOI: 10.1080/21580103.2012.658235

Sukhadiya, M., C.A. Dholariya, L.K. Behera, A.A. Mehta, D.P. Patel and R.P. Gunaga. 2019. Commercial utilization and propagation of gum karaya: *Sterculia urens* Roxb. *MFP News* 29(3): 5–8.

Tavares, W.S., M.F., Souza-Filho, E.L. Araujo, J.E. Serrão and J.C. Zanuncio. 2013. Ecological observations and germination of *Sterculia chicha* Seeds Colonized by *Anastrepha bezzii. International Journal of Environmental Research* 7(3): 795–800.

8 In Vitro Propagation of *Sterculia urens* Species

T. Pullaiah and Mallappa Kumara Swamy

8.1 INTRODUCTION

Sterculia urens, an intermediate tree, produces an acetylated polysaccharide exudate and is commonly branded as gum karaya. It is widely used in pharma industries as an emulsifier, stabilizer, and thickener. Gum karaya is also used in petroleum products and in the textile industry. In food and dairy industries, it is majorly used as an additive agent in the preparation of ice creams, sauces, and beverages, owing to the gum acting as an emulsifier and thickener (Gautami and Bhat, 1992; Subhashini Devi et al., 2010). In nature, the propagation of *S. urens* occurs via seeds. Nevertheless, the viability of seeds is seen to be in decline as time progresses. Furthermore, the propagation of *S. urens* by means of stem cutting method includes several difficulties and challenges. *S. urens* population is observed to be depleting in their natural habitat due to destructive harvesting methods, overexploitation, and low seedling survival rate (Purohit and Dave, 1996; Sunnichan et al., 1998; Hussain et al., 2007). Commercially, tapping/harvesting of gum karaya is accomplished by means of peeling, blazing, and creating penetrative cuts. Thus, these approaches cause injury to trees and can result in trees loss. Overexploitation by blazing too deeply and exposing wood to borer attacks are key reasons for the destruction of these trees.

The low propagation rate, combined with overexploitation and stumpy seedlings survival rate, has significantly caused a reduction in *S. urens* species population in its natural habitat. This has widened the gap between demand and supply for gum karaya, thus impacting pressure on the species. The government of Madhya Pradesh, India, has taken measures by imposing banning on tapping and gathering of *S. urens* tree exudates to permit retrieval and regeneration of trees (Nair et al., 1995). Even then, there is a serious worry about the loss of wild *S. urens* germplasm. Considering the high demand for gum karaya, large-scale cultivation of *S. urens* is needed. Due to lower rates of natural propagation, *ex situ* conservation approaches should be involved. In vitro propagation technique is one of the best ways for propagating any plant species within a short time. The following sections will describe different in vitro approaches adopted for the propagation of *S. urens* species.

DOI: 10.1201/9781003438960-8

8.2 MICROPROPAGATION OF *STERCULIA URENS*

8.2.1 Shoot Induction

Purohit and Dave (1996) used the cotyledonary nodes of seedling explant and reported shoot induction. They placed the explants on MS media (Murashige and Skoog, 1962) supplemented with 2 mg/l BAP (6-Benzylaminopurine). Bahadur and Shailaja (1997) and Shailaja et al. (1997) reported shoot formation using seedling shoot tips and nodal segments as explants. Sunnichan et al. (1998) used mature nodal segments as explants source and reported the formation of six shoots on average when cultured on MS media containing 6.62μM BAP. These shoots showed rooting after transferring to 1/4 MS media having 9.82 μM IBA (Indole-3-butyric). They reported nodular calli development from hypocotyl explants cultured on MS media consisting of 4.52 μM 2,4-D (2,4-Dichlorophenoxyacetic acid) and 8.90 μM BAP. The formation of somatic embryos from the nodular callus was observed when transferred to media having 0.45 μM TDZ (Thidiazuron) for two days. Nearly 30% of somatic embryos formed into plantlets after placing onto 1/4 MS basal media devoid of plant growth regulators. Anitha (1997) and Anitha and Pullaiah (1999) reported that 2 mg/l BAP was suitable concentration for shoot regeneration. They studied the influence of coconut milk as additive to the shoot improvement, but the result was not encouraging. In callus induction studies carried out by them, a profuse callus formation was witnessed from cotyledonary explants placed on 2 mg/l 2,4-D supplemented medium. In a study by Subhashini Devi et al. (2010), the healthy seeds were chosen for studying the in vitro regeneration capacity of seeds. Initially, they treated seeds with sulfuric acid (concentrated) for 1 min and later washed cautiously under running tap water. This was followed by treatment with Teepol (5% (w/v)) for 10 min, and later with bavistin, a commercial fungicide, for 5 min. The seeds were successively surface-sterilized with $HgCl_2$ (0.1% (w/v)) for 5 min and thoroughly washed with sterilized distilled water and soaked in distilled water for 1 day. Before inoculating, the seeds were surface-dried on sterile filter papers and cultured on MS medium. After 30 days of culturing, in vitro raised plants showed approximately 3–4 nodes. Hussain et al. (2007) reported the highest shoot multiplication rate using cotyledonary node only, when placed on 2.27μM TDZ and ascorbic acid (0.1%) supplemented medium. Shoot cultures were established by them via repeatedly sub-culturing the original cotyledonary nodes on media supplemented with 0.45 μM TDZ, even after the second harvest. Hussain et al. (2008) described a reproducible procedure for the micropropagation of *S. urens* by using intact seedlings as source of explants on MS media consisting of 5.0 μM TDZ and 1.5 μM GA_3 (gibberellic acid) and ascorbic acid (0.1%). They observed the formation of the highest number of shoots (14) within 56 days of seed culturing on MS medium, and root formation was found to be absent. They established proliferative shoot cultures through repetitive sub-culturing of mother seedlings on fresh media, viz. MS media with 2.5 μM TDZ, 1.5 μM GA_3, and ascorbic acid (0.1%), after removing all freshly developed shoots. They also stated that the shoot-forming ability of seeds is affected by the types and concentrations of cytokinins used in the media.

Subhashini Devi et al. (2010, 2017a) and Subhashini Devi (2017a) reported micropropagation of *S. urens* though nodal cultures. According to them, MS medium fortified with 0.90μM TDZ and 13.33μM BAP gave the best results. Polyphenol oxidase, peroxidase, total proteins, total carbohydrates, and total phenols were biochemically analyzed at an interval of 20, 40, and 60 days from the micropropagated plants. The results showed that biochemical parameters observed in the in vitro derived plants remained similar to that of the mother plants.

Subhashini Devi et al. (2011) and Subhashini Devi (2017a) reported the formation of increased number of shoots (7.5) from nodal explants obtained from 30-day-old in vitro raised plants on MS media having 13.3 μM BAP, while on MS media containing 0.9 μM TDZ, about 8.68 shoots were noticed. Addition of either one of BAP or TDZ shows higher number of shoots in comparison to previously reported data on regeneration from mature trees (Figure 8.1). It advocates that the explant

FIGURE 8.1 Different stages of micropropagation of *Sterculia urens*: (a) multiple shoot induction on MS + 0.90 μM TDZ from nodal explants of 1-month-old plants (Bar = 0.5 cm); (b) multiple shoot induction on MS + 0.90 μM TDZ from cotyledonary node explants of 15-day-old seedling (Bar = 0.5 cm); (c) elongation of shoots on MS + 8.7 μM GA_3 after 30 days (Bar = 2 cm); (d) in vitro rooting on half-strength MS + 19.70 μM IBA and half-strength MS + 16.12 μM NAA (Bar = 2.5 cm); (e) acclimatization of *Sterculia urens* (3 months old) (Bar = 8 cm).

Source: Subhashini Devi et al. (2011).

sources obtained from juvenile plants have higher regeneration capability as compared to explants isolated from mature trees. This could be attributed to the fact that adult trees exhibit less differentiation rate and juvenile tissues occurring in seedling explants exhibit higher differentiation rate. Though multiple shoots were encouraged on a TDZ-containing media, the elongation of shoots was found to be absent. Thus, for elongation of shoots, MS media consisting of 8.7 μM GA_3 was used.

Earlier findings by Purohit and Dave (1996) reported that on 8.9 μM BAP supplemented MS media, around 16 shoots were formed per cotyledonary node explants in 3 harvests over a period of 9 weeks. Likewise, Hussain et al. (2007) showed a maximum number of 19 shoots per cotyledonary node explants on 2.27 μM TDZ and ascorbic acid (0.1%) supplemented MS media after 8 weeks. Subhashini Devi et al. (2011) recorded the formation of 11.24 shoots per cotyledonary node explants in a single harvest on 0.9 μM TDZ contained MS media within 35 days. According to all these researchers' findings, it is suggested that cotyledonary node explants of *S. urens* possess higher regenerative capability as compared to nodal segments. This may possibly be due to the fact that the cotyledonary nodes have more embryonic potential. Explant types and their growth phase can significantly influence multiple shoots induction in many tree species.

Subhashini Devi et al. (2013) and Subhshini Devi (2017b) cultured cotyledonary node explants on MS media containing 0.5 mg/l BAP and 4 mg/l NAA (naphthaleneacetic acid). They observed the induction and establishment of calli within 1 week. Further, the lower concentrations of NAA resulted to a regenerating callus (RC), while non-regenerating callus (NRC) was noticed when higher levels of NAA were used in the media. In the RC, increased activities of peroxidases (POD), catalase (CAT), and superoxide dismutase (SOD) were detected, whereas in the NRC, decreased enzyme activities were recorded. Overall, in calli, six POD isoenzymes were identified. In the NRC, the protein content was observed to be less, and it improved considerably during the regeneration of shoots from calli. However, the total phenols and flavonoids contents were higher in the NRC. Data from sodium dodecyl sulphate-polyacrylamide gel electrophoresis (SDS-PAGE) analysis exposed the roles of several proteins in organogenesis. During in vitro regeneration of *S. urens*, oxidative stress was developed, and this could be because of lesser activities of POD, CAT, and SOD. Overall, there exists a direct link between enhanced activities of antioxidant enzymes and the occurrence of reactive oxygen species (ROS) in *S. urens*. The increased activities of POD, CAT, and SOD in RC as compared to NRC suggest the effective ROS (including H_2O_2) scavenging activities in regenerated shoots. There occurs a fine regulation of ROS during organogenesis of shoots in *S. urens*, as revealed by the specific POD isoforms in both RC and NRC.

Dhiman et al. (2020) cultured cotyledonary nodes on MS media added with 0.33 mg/l TDZ and found greater regeneration response (81.9%) with maximum shoot numbers (19.62) after 35 days of incubation. For the elongation of shoots, in vitro regenerated shoots were cultured on MS media supplemented with a combination of plant growth regulators, namely, 0.2 mg/l BAP, 1.5 mg/l Kn, and 2.0 mg/l GA_3. On this elongation media, optimum shoot elongation up to 5 cm was recorded within 14 days. Addition of 1.0 mg/l $AgNO_3$ along with 1.0 mg/l Kn (kinetin) to MS media produced considerable shoot quality with dark-green and shiny leaves. Also, these

TABLE 8.1
Micropropagation studies in *Sterculia urens*

Explant	Medium + plant growth regulators	Result	Reference
Cotyledonary node	MS + 2 mg/l BAP	Shoots	Purohit and Dave, 1996
Seedling shoot apex		Multiple shoots	Bahadur and Shailaja, 1997
Node		Multiple shoots	Shailaja et al., 1997
Mature tree node	MS + 6.62 μM BAP	6 shoots per node, somatic embryos	Sunnichan et al., 1998
Cotyledonary node	+ 2 mg/l BAP	Shoots, plantlets, callus	Anitha, 1997; Anitha and Pullaiah, 1999
Cotyledonary node	2.27 μM TDZ + 0.1% ascorbic acid	Adventitious shoots, plantlets	Hussain et al., 2007
Seedling	MS + 5.0 μM TDZ + 1.5 μM GA_3 + 0.1% ascorbic acid	14 shoots from seedling	Hussain et al., 2008
Nodes from seedlings	MS + 0.90 μM TDZ + 13.33 μM BAP	8.68 shoots per explant	Subhashini Devi et al., 2010
Cotyledonary node and node	MS + 0.9 μM TDZ	8.6 shoots per node and 11.2 shoots per cotyledonary node	Subhashini Devi et al., 2011
Cotyledonary node	MS+0.5 mg dm^{-3} BAP + 2 or 4 mg dm^{-3} NAA	Callus induction and organogenesis	Subhashini Devi et al., 2013
Nodal explants	Synthetic seeds; MS + TDZ	Regeneration potential 73%	Subhashini Devi et al., 2014
Cotyledonary node	MS + TDZ	81.9% regeneration response	Dhiman et al., 2020

treatments encouraged the growth of stems, improved stem strength, abridged internodal lengths (de-etiolation), and restricted the development of callus.

8.2.2 Rooting and Acclimatization

In general, the shoots, measuring 2 to 3 cm in height, are procured after the elongation phase and thoroughly washed under running tap water so as to get rid of agar and moved to rooting medium for rooting. Purohit and Dave (1996) transferred the elongated shoots to 1/4 MS media after pulse treatment with 500 mg/l IBA for 10 min and achieved better root induction. Anitha and Pullaiah (1999) rooted the shoots on 1/2 MS medium containing 1 mg/l IBA (indole-3-butyric acid). Hussain et al. (2007) rooted shoots on 1/4 MS media supplemented with 9.80 μM IBA for 8 weeks and later acclimatized the plants using autoclaved vermiculite. Likewise, Hussain et al. (2008) also recommended the use of 9.80 μM IBA in 1/4 MS media for better rooting. Subhashini Devi et al. (2010) reported successful rooting on 1/4 MS media having 19.70 μM IBA or 16.12 μM NAA within 10 weeks. Also, Subhashini Devi et al. (2011) stated that rooting can be better achieved on 1/2 strength MS media also; however, this needs the addition of either 19.7 μM IBA or 16.12 μM NAA to the media. Similarly, the use of 1/2 strength MS media was justified by Dhiman et al. (2020). According to them, for an ideal rhizogenesis response, shoots are to be given a pulse treatment with IBA (200 mg/l) for 3–4 min, followed by culturing on rooting media, that is, 1/2 MS media fortified with 2.0 mg/l IBA. Further, a higher plant survival rate (up to 80%) was observed when biotization using mycorrhizae (*Glomus* sp.) was done at the hardening stage. Plants treated with *Glomus* sp. improved the root system and showed healthy survival, with dark-green-coloured leaves. Also, the plants were less-susceptible to diseases/infections when compared to untreated plants. The use of molecular markers revealed the existence of 95% genetic homogeneity amongst the tissue culture raised and mother plants. Plantlets with well-developed roots are detached and moved to plastic cups having garden soil (autoclaved) mixed with vermiculite and sand (1:1:1). They were kept covered with polythene bags with holes for maintaining up to 70% humidity and watered every alternative days with Hoagland and Arnon (1950) solution and placed in a culture room environment until the new leaves start to appear. Later, polythene bags are uncovered and kept under greenhouse settings for about 28 days. As soon as the plantlets get adjusted to greenhouse conditions, they are transferred to regular field conditions.

As specified by Subhashini Devi et al. (2010), in vitro derived plants exhibit normal growth and development. The biochemical analysis has revealed the secretion of important compounds required for plants' growth, nutrition, and superior resistance to pathogenic organisms. They established that micropropagated *S. urens* plants display improved levels of enzymatic activities and constitute higher contents of total proteins, total carbohydrates, and phenols as compared to control plants and therefore exhibit greater percentage of survival rate. However, micropropagated plants showed lesser percentage of field establishments, and this could be because of various reasons. One of the reasons may be altered biochemical compositions.

8.2.2.1 Synthetic Seed Technology

Subhashini Devi et al. (2014) successfully encapsulated nodal segments of *S. urens* and showed plant regeneration prospect. They have chosen alginate for encapsulation as it displays rapid gelation, has less toxicity to plant tissues, improves capsule-forming capability, and also the alginate beads rigidity safeguards the enclosed micropropagules against mechanical injuries. The use of sodium alginate (4%) upon complexation with $CaCl_2.2H_2O$ (100 mM) for 30 min produced very clear, firm, isodiametric beads of sustainable, identical size and shape. Sodium alginate concentrations lower than 4% formed beads that were too soft and not appropriate to handle; however, at higher concentrations (>6%), beads were very hard, viscous, and stalled the development of shoots. For evaluating the regeneration potential of encapsulated nodal segments, different growth media were used, namely, MS basal media, MS media with 3.0 mg/l BAP, MS media with 0.2 mg/l TDZ, and MS media with 0.5 mg/l BAP, and 2.0 mg/l NAA. The regeneration frequency was found to be the highest (73.33%) on MS media containing 0.2 mg/l TDZ, followed by 69.33% regeneration frequency on MS media supplemented with 3 mg/l BAP after 1 week (Figure 8.2). The regeneration potential was noticed even

FIGURE 8.2 (a) Effect of 2, 4, and 6% sodium alginate on encapsulation and bead formation; (b) shoot emergence from encapsulated nodal explants on MS basal medium; and (c) shoot emergence from encapsulated nodal explants on MS basal medium supplemented with growth regulators (0.2 mg/l TDZ and 3 mg/l BA).

Source: Subhashini Devi et al. (2014).

after six months of storing at 4°C. These results suggested the possibility of using synthetic seed technology as an alternative approach for germplasm conservation of *S. urens*.

8.3 CONCLUSIONS AND FUTURE PROSPECTS

Plant micropropagation is one of the potent techniques for obtaining plant products with several industrial uses other than using whole plants. In vitro propagation is a valuable approach, as many plant secondary metabolites cannot be synthesised by chemical methods. Overexploitation has suggestively led to reduced population of *S. urens* in nature. Due to the increased demand for gum karaya, there is a high pressure on this species. Thus, the large-scale cultivation of *S. urens* must be encouraged. Owing to drawbacks of natural propagation, *ex situ* conservation approaches should be involved. Thus, in vitro propagation of *S. urens* is the best ways for large-scale production in a short amount of time. Research findings have suggested using cotyledonary nodes as the best explant. MS media supplemented with BAP either alone or in combination with TDZ encouraged in vitro shoot regeneration. Addition of 1.5 μM GA_3 and ascorbic acid (0.1%) to MS media increased shoot length. However, when media was supplemented with 0.5 mg/l BAP and 4 mg/l NAA, callus formation was induced. For better rooting, 19.7 μM IBA or 16.12 μM NAA added to 1/2 strength MS media suits the best. Biotization using mycorrhizae (*Glomus* sp.) at the hardening stage encouraged higher rate of plant survival in the field conditions. Synthetic seed technology can be utilized in conserving *S. urens* germplasms. Biotechnological tactics allied with plant tissue culture increase the opportunity of exploring medicinal plants and their bioactive metabolites, along with traditional agriculture. In the coming years, tissue culture approach should realize its full prospects, considering the involvement of novel technologies, such as cell culture, gene manipulation, gene editing, metabolic pathways exploration, etc. However, the main limitations must be addressed for tolerable profit-making applications of *in vitro* regenerated *S. urens* plants on a large scale.

REFERENCES

Anitha, S. 1997. *In vitro* propagation studies of *Sterculia urens*. M.Phil. dissertation, Sri Krishnadevaraya University, Anantapur.

Anitha, S. and T. Pullaiah. 1999. *In vitro* propagation studies of *Sterculia urens* Roxb. In: Kavi Kishor, P.B. (ed.) *Emerging trends in plant tissue culture and molecular biology*. Universities Press, Hyderabad, pp. 146–150.

Bahadur, B. and A. Shailaja. 1997. Multiple shoot formation from shoot apex of karaya gum, *Sterculia urens* Roxb. *J. Swamy Bot. Cl.* 14: 114–116.

Dhiman, M., L. Sharma, A. Singh and M.M. Sharma. 2020. *Ex situ* conservation using *in vitro* methods of an endangered plant *Sterculia urens* Roxb.: A high volume trade plant for gum karaya. *Indus. Crops Prod.* 158: Article 113015.

Gautami, S. and R.V. Bhat. 1992. A monograph of gum karaya. National Institute of Nutrition, ICMR, Hyderabad, India.

Hoagland, D.R. and D.I. Arnon. 1950. The water-culture method for growing plants without soil. (Circular (California Agricultural Experiment Station), 347. ed.). University of California, College of Agriculture, Agricultural Experiment Station, Berkeley, CA.

Hussain, T.M., T. Chandrasekhar and G.R. Gopal. 2007. High frequency shoot regeneration of *Sterculia urens* Roxb., an endangered tree species through cotyledonary node cultures. *African J. Biotech.* 6: 1643–1649.

Hussain, T.M., T. Chandrasekhar and G.R. Gopal. 2008 Micropropagation of *Sterculia urens* Roxb. an endangered tree species from intact seedlings. *African J. Biotech.* 7(2): 95–101.

Murashige, T. and F. Skoog. 1962. A revised medium for rapid growth and bioassays with tobacco tissue culture. *Physiol. Plant.* 15: 473–497.

Nair, M.N.B., K.R. Shivanna and H.Y. Mohan Ram. 1995. Ethephon enhances Karaya gum yield and wound healing response: a preliminary report. *Curr. Sci.* 69: 809–810.

Purohit, S.D. and A. Dave. 1996. Micropropagation of *Sterculia urens* Roxb., an endangered tree species. *Plant Cell Rep.* 15: 704–706.

Shailaja, A., D. Laxmi, E. Chamundeswari and B. Bahadur. 1997. Multiple shoot induction from nodal explants of *Sterculia urens* Roxb. And important gum yielding tree. *J. Swamy Bot. Cl.* 14: 98–100.

Subhashini Devi, P. 2017a. Propagation of *Sterculia urens* Roxb. using tissue culture approaches. Lambert Academic Publishing, London.

Subhashini Devi, P. 2017b. Antioxidants and their role in micropropagation of *Sterculia urens* Roxb. Lambert Academic Publishing, London.

Subhashini Devi, P., A. Arundhati and T. Raghavarao. 2011. Multiple shoot induction and regeneration of whole plants from cotyledonary node and nodal explants of *Sterculia urens* Roxb., a gum yielding tree. *J. Plant. Biochem. Biotech.* 20(2): 161–165. DOI: 10.1007/s13562-011-0041-x

Subhashini Devi, P., B. Satyanarayana, A. Arundathi and T. Raghava Rao. 2010. Biochemical analysis of in vitro derived plants of *Sterculia urens* during acclimatization. *J. Chem. Pharmaceut. Res.* 2(4): 635–641.

Subhashini Devi, P., B. Satyanarayana, A. Arundhati and T. Raghavarao. 2013. Activity of antioxidant enzymes and secondary metabolites during *in vitro* regeneration of *Sterculia urens*. *Biologia Plantarum* 57(4). DOI: 10.1007/s10535-013-0337-x

Subhashini Devi, P., B. Satyanarayana, A. Arundhati and T. Raghavarao. 2014. Plant regeneration from encapsulated nodal segments of *Sterculia urens* Roxb., an endangered gum-yielding tree. *J. Crop Sci. Biotech* 17(1): 21–25.

Sunnichan, V.G., K.R. Shivanna and H.Y. Mohan Ram. 1998. Micropropagation of gum karaya (*Sterculia urens*) by adventitious shoot formation and somatic embryogenesis. *Plant Cell Rep.* 17: 951–956.

9 Assessment of Genetic Diversity in *Sterculia* Species Using Molecular Studies

Mallappa Kumara Swamy,
Boregowda Purushotham,
Abhijit Dey, and Ajay Kumar

9.1 INTRODUCTION

Plants have become the major resource for the development of most of the products/ingredients that common people consume in their daily life, including nourishment, housing, medications, apparel, etc. At present, human beings are totally dependent on indigenous plant species in one way or another. Thus, conservation of a flora ecosystem is of prime importance and should not be neglected. Conservation ecology includes the scientific management of natural ecosystem to uplift the biodiversity and to protect wild species, particularly in the forests (Jaisankar et al., 2018). Safeguarding of biodiversity is very imperative for a sustainable future. Plant species provide numerous advantages to human beings, and thus, protecting the endangered biota is of great importance (Maxted, 2001; Nath et al., 2023). A better future can be expected with the expansion of species biodiversity, preserving ecosystem, and genetic diversity conservation. In this regard, understanding on genetic variations within plant species and population is very vital. In order to improve the conservation status of any species, their genetic diversity should be conserved. There is a possibility of introducing desired genetic modifications via plant breeding programs. Also, genetic diversity conservation helps in understanding the possible threats which could result in species loss and in implementing strategies to reduce the threats (Plenk et al., 2019). Plant breeding is a way of conserving biodiversity, and it is used for introducing a desired genetic modification in plant species. Plant breeding mainly involves three mechanisms: (1) creating genetic variations, (2) screening and selection, and (3) facilitating them for commercial viability. It is majorly utilized for the improvement of crops by allowing the insertion of superior alleles in the gene pool (Borrelli et al., 2009; Nath et al., 2023).

Low breeding and genetic diversity could adversely upset the fitness of species population (Jiménez et al., 2014). Previously, several investigations have indicated

 DOI: 10.1201/9781003438960-9

that the extinction of species is largely triggered by lower genetic diversity, causing a reduction in the population growth and survival rate (Jiménez et al., 2014; Plenk et al., 2019). In addition, external reasons, such as anthropogenic disturbances, ecological stresses, drought responses, atmospheric CO_2 level, influence stomatal density and chlorophyll content as an adaptive response of plants (Dunn et al., 2019; Spinedi et al., 2019). Therefore, it is imperative to assess the genetic difference and anatomical features connected with plant growth and survival for species breeding and conservation.

In addition to commercial values, forest tree species have environmental prominence in all ecosystems. Forest trees are mostly found in nature and narrowly domesticated or selected artificially (Savolainen and Pyhäjärvi, 2007). In forest tree improvement programs, in addition to morphological and physiological features, it is required to preserve and explore the genetic diversity of the population. These programs may improve the genetic values of both native species and exotic species. Also, to restore a forest, tree breeding programs provide a superior quality seeds from approved sources of propagules and/or seeds associated with silviculture method (Bell et al., 2008). Though it is labour-demanding and costly, improvement of genotypes can be expected. *Sterculia* species, belonging to the family Sterculiaceae, is currently placed under Malvaceae, well known for their commercial importance, and possesses significant ethnobotanical, pharmacological applications (Alverson et al., 1999; Galla and Dubasi, 2010; Luján-Medina et al., 2013; Dhiman et al., 2019; Prasad et al., 2022, 2023). They are also recognized as cacao family members and are distributed all over the world, mainly concentrated in the tropical and sub-tropical regions (Al-Rahman, 2017; Dhiman et al., 2019). *Sterculia* species release gum exudates in response to wounds on the tissue parts as a defence mechanism. The gum exudate, also known as gum karaya, is one of the most inexpensive and richest natural raw material and widely acknowledged for its numerous commercial applications. Consequently, these tree species are unscientifically tapped and expansively harvested for their bark, leading to species threat (Saragih and Siswadi, 2019; Jannah et al., 2022). Currently, a few of the *Sterculia* species like, *S. urens* and *S. quadrifida*, are facing an endangered status, warranting their conservation. Therefore, efforts are needed immediately for establishing fundamental information on *Sterculia* species. Various steps required to be taken to provide genetic material in the process of tree improvement programs comprise exploration, preservation, and assessment of the characters of the plant species so that the information is largely known. Thus, investigations on breeding, preservation, genetic diversity, morphological features, physiological characteristics, and anatomical features associated with growth, development, and survival are critical for conserving *Sterculia* species. This chapter aims to provide comprehensive data on genetic diversity and molecular-level understanding of different species of *Sterculia*.

9.2 A BRIEF NOTE ON MOLECULAR MARKERS

Some of the collective methodologies, such as molecular markers, bioinformatics, and geographical information systems, help develop appropriate procedures for assessing plant genomic diversities (Indu et al., 2019). Molecular markers (also known as DNA-based markers or genetic markers) are DNA segments linked to a specific region within the genome. These are utilized to identify a specific DNA fragment in a collection of unknown DNAs. DNA polymorphism results due to any one of the following reasons, viz., additions, deletions, duplications, point mutations, and translocations in genomes. However, these polymorphic changes may not certainly upset the genes' functions (Indu et al., 2019). Various molecular markers are used to establish genomic diversity levels. They are relatively specific and can differentiate plants from the phenological differences. As compared to morphological approaches, molecular marker techniques are more accurate and thus widely used (Swamy and Anuradha, 2011; Nath et al., 2023). Moreover, these molecular markers help in developing more reliable and effective tactics while selecting and in the management of numerous plant species (Jaisankar et al., 2017). Molecular marker techniques are categorized into three sets on the basis of their detection: (1) hybridization-based, (2) PCR-based (or polymerase chain reaction–based), and (3) DNA sequence–based (Nikmah et al., 2016; Wahyudi et al., 2020; Jannah et al., 2022). Largely used molecular marker approaches include restriction fragment length polymorphisms (RFLPs), random amplified polymorphic DNAs (RAPDs), inter simple sequence repeat (ISSR), amplified fragment length polymorphism (AFLP), single sequence repeats (SSRs), and single-stranded conformation polymorphism (SSCP). Whole-genome sequences (WGS) are amplified and later separated through electrophoresis approach. They are presently used for studies related to the assessment of biodiversity, germplasm analysis, and the construction of gene maps. It is noteworthy to mention that it needs a very small quantity of DNA (Kloch et al., 2015; Nath et al., 2023).

With the advent of next-generation sequencing (NGS) technologies, there is a possibility of whole genome sequencing or RNA-seq transcriptome of a target plant readily and economically (Lee et al., 2015; Eum et al., 2019). These novel methods have allowed scientists to understand genomic knowledge on both model and non-model plant species, including several medicinally valued plants. Simple sequence length polymorphism (SSLP) is utilized for mapping single loci, and also for fingerprinting in plants. There can be monomeric, dimeric, trimeric, and tetrameric motif of a repetitive sequence. Single-nucleotide polymorphisms (SNPs) are used for identifying genetic variations due to mutation (Mahajan and Gupta, 2012; Huang et al., 2010; Huang et al., 2018; Eum et al., 2019). Overall, these approaches are employed for the authentication of species or varieties, understanding genetic diversity status, to establish phylogenetic correlations, and also for gene mapping drives (Lidder and Sonnino, 2012; Govindaraj et al., 2015; Poerba et al., 2018; Jaisankar et al., 2018; Uslan and Pharmawati, 2020; Wahyudi et al., 2020; Hermenegildo et al., 2021; Jannah et al., 2022). The following section discusses various molecular marker studies involved in different *Sterculia* species.

9.3 MOLECULAR MARKER STUDIES IN *STERCULIA* SPECIES

The genetic diversity of *S. quadrifida* plants, occurring in eight different subdistricts in Kupang, Indonesia, was investigated by Uslan and Pharmawati, (2020). They isolated DNA samples and amplified with six RAPD primers, namely, OPF-11, OPD-11, OPB-04, UBC-106, UBC-127, and UBC-250. They recorded a total of 131 DNA fragments that ranged between 250 bp and 1,400 bp. Unweighted pair group method with arithmetic mean (UPGMA) approach and Nei and Li similarity coefficient study showed the grouping of *S. quadrifida* populations into two major clusters and 12 subclusters. The *S. quadrifida* population from Fatuleu region showed the lowest genetic diversity (0.0305), while the highest genetic diversity (0.1050) was noticed in populations obtained from Kelapa Lima region. Overall, their study suggested the existence of high genetic diversity amongst the population of *S. quadrifida* in Kupang area. Also, the populations were found to cluster on the basis of their topographical distribution.

To identify genetic diversity in *S. quadrifida* occurring in Kupang, Indonesia, ISSR markers were used by Jannah et al. (2022). Also, they evaluated the stomatal density and chlorophyll content in plants collected from different vicinities. Interestingly, to analyze the genetic diversity, they collected samples from the Oelsonbai forest area, Kupang, while samples obtained from regions with varying degrees of anthropogenic activities in Kupang City were used for analyzing stomatal density and chlorophyll content. The isolated DNA was amplified through the use of UBC 811 (5' 3'; GAG AGA GAG AGA GAG AC), UBC 812 (5' 3'; GAG AGA GAG AGA GAG AA) and UBC 815 (5' 3'; CTC TCT CTC TCT CTC TG) ISSR primers. Stomatal density was detected by means of a microscopic examination, and a spectrophotometry method was used for determining chlorophyll content. The results revealed the occurrence of a total of 78 ISSR bands that ranged between 320 bp and 3,200 bp (Figure 9.1). DNA polymorphism is the difference observed in DNA sequences amongst individual plants, groups, or population. Thus, to identify the polymorphism, the obtained genetic data was subjected to UPGMA method and Nei and Li coefficient similarity study. The resultant phylogenetic tree showed the population with four major clusters and six subclusters (Figure 9.2). There was a moderate value of heterozygosity (0.5), suggesting the unassertive genetic diversity of *S. quadrifida*. Interestingly, their study suggested that *S. quadrifida* samples collected from the Oelsonbai forest region were not clustered geologically; however, both isolated and adjacent *S. quadrifida* populations in the same region tend to cluster together. Overall, these results advocated the fact that geographical distribution was not linked with genetic distance. This was attributed to special effects of environmental factors, varying plants' age, or plants' cross-breeding. Further, dissimilarity in the chlorophyll content and stomatal density was noticed, and they predicted this disparity to anthropogenic influences.

The plant species of *Sterculia* genus with high prospective for identifying novel drugs/lead molecules lack genomic information and thus hinder the validation of their genetic background for future usage. The advancement of NGS technologies has enabled sequencing of the whole genome or RNA transcripts cost-effectively, even in tree species. RNA-seq transcriptome is used for evaluating genomics of medicinal plants in recent times, and it is widely growing. It is one of the molecular tools to dissect basic metabolic pathways of unknown medicinally valued plant species and to recognize lead molecules for discovering drugs (Bae et al., 2018). However, genomic understanding of *Sterculia* species is yet to be fully explored. Considering this, for the first time, Eum et al. (2019) performed RNA-seq transcriptome analysis for obtaining genomic resources in *S. lanceolata*. RNA-seq resulting assembled unigenes were utilized for analyzing gene ontology (GO), searching Kyoto Encyclopedia of Genes and

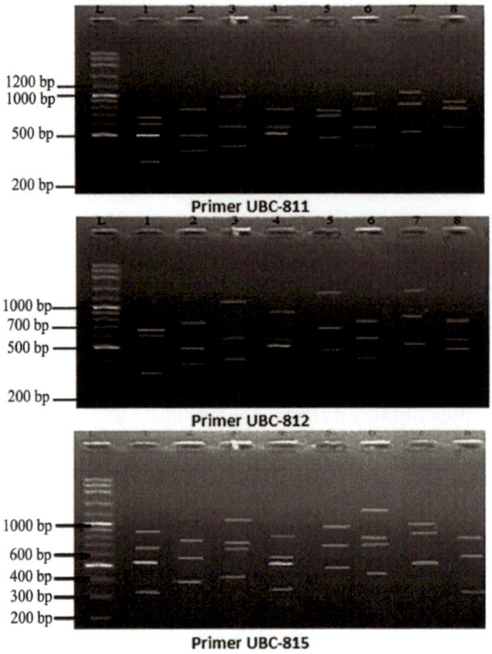

FIGURE 9.1 ISSR profile of each primer (UBC-811, UBC-812, and UBC 815) on *Sterculia quadrifida* from Oelsonbai forest area, Kupang. Note: 1–2, Talaka; 3–4, Sokon; 5–6, Naioni; 7–8, Sikumana. DNA ladder VC 100 bp Plus (L).

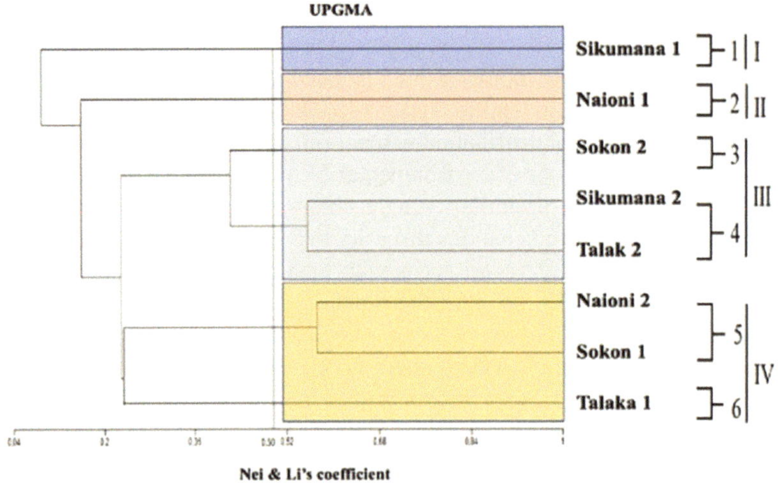

FIGURE 9.2 Dendrogram of eight samples of *Sterculia quadrifida* from Oelsonbai forest area, Kupang, based on Nei and Li's coefficient similarity with UPGMA method in MVSP software. Note: Arabic number denotes subcluster; roman number denotes cluster.

Genomes (KEGG) metabolic pathway, and mining SSR markers. The polymorphic data obtained from SSR markers were used for examining genetic resemblances amongst 15 species of *Sterculia*. RNA-seq data resulted overall to 28,088,794 clear reads having 3,692,930,973 bp of length. By using de novo assembling, 7,857 (3,559,905 bp) non-redundant unigenes were generated, and each unigene had an approximate length of 453 bp. Out of 1,893 unigenes allocated by the KEGG Orthology database, nearly 1,158 unigenes were linked to 328 different metabolic pathways. About 159 unigenes were linked to the biosynthesis of secondary metabolites pathways. Among them, 72 unigenes are known to take part in the biosynthesis of terpenoids, phenylpropanoids, and flavonoids. A total of 899 SSR were recognized, and 19 polymorphic SSR markers were established. By using allelic data, both principal coordinates analysis and neighbour joining analysis produced three clusters amongst 15 species of *Sterculia*. Overall, these results provided a valuable resource for genomic investigations to untangle primary mechanisms involved in therapeutic activities of different species of *Sterculia*. Moreover, this reference transcriptomic information would be a very advantageous resource to facilitate genetic marker discoveries and other genomic, transcriptomic, metabolomics, and biological investigations of *Sterculia* species.

9.4 CONCLUSIONS

As pronounced, most species of *Sterculia* have abundant commercial/pharmaceutical importance. Nevertheless, very poor genomic or genetic researches have been performed as of now, and molecular data for the species belonging to this genus is yet to be established well. Genomic resource is very essential to develop an appropriate DNA marker for identifying or authenticating species having pharmaceutical applications. There is a limited molecular information available to researchers, and this could be because of the fact that most of the research work on this genus is focused towards investigating pharmaceutical effects rather than molecular studies. The molecular markers are unquestionably precious tools to address plant genetics and breeding problems. To date, RAPD and ISSR markers have been explored to examine genetic diversity in a few *Sterculia* species. There is only one attempt made to understand the genomic information of *S. lanceolata* using advanced NGS approaches. Therefore, more number of DNA-based genetic markers should be explored to examine the genetic diversity of *Sterculia* species in the near future. Molecular data on *Sterculia* species will be very useful during the identification, selection, authentication, preservation, and genetic variation understanding. In total, molecular-level data can assist in *Sterculia* species breeding programmes. Moreover, these genetic markers act as a promising tool in overcoming the problems associated with *Sterculia* plant products adulteration.

REFERENCES

Al-Rahman, A.Y.R. 2017. A Pharmacognostical Study of Certain *Sterculia* species (Family: Sterculiaceae). Cairo University, Giza. PhD Thesis.

Alverson, W.S., B.A. Whitlock, R. Nyffeler, C. Bayer and D.A. Baum. 1999. Phylogeny of the core Malvales: evidence from ndhF sequence data. *Am. J. Bot.* 86: 1474–1486. https://doi.org/10.2307/2656928.

Bae, D.Y., S.M. Eum, S.W. Lee, J.H. Paik, S.Y. Kim, M. Park, C. Lee, T.B. Tran, V.H. Do, J.Y. Heo, E.S. Seong, I.S. Kim, K.Y. Choi, J.S. Hong, R.V. Ramekar, S. Choi and J.K. Na. 2018. Enrichment of genomic resources and identification of simple sequence repeats from medicinally important *Clausena excavata*. *3 Biotech*. 8(133): 1. https://doi.org/10.1007/s13205–018–1162-x

Bell, F.W., J. Parton, N. Stocker, D. Joyce, D. Reid, M. Wester, A. Stinson, G. Kayahara and B. Towill. 2008. Developing a silvicultural framework and definitions for use in forest management planning and practice. *For. Chronicles*. 84(5): 678–693.

Borrelli, G.M., L. Orru, P.D. Vita, D. Barabaschi, A.M. Mastrangelo and L. Cattivelli. 2009. Integrated views in plant breeding: Modern approaches for an old topic. In: Sadras, V., Calderini, D.F. (eds.) *Crop Physiology*. Elsevier Inc., New York.

Dhiman, M., A. Singh and M.M. Sharma. 2019. A review on *Sterculia urens* Roxb.: A boon to the livelihood for tribal people and industry. *Ind. Crops. Prod.* 130: 341–351.

Dunn, J., L. Hunt, M. Afsharinafar, M.A. Meselmani, A. Mitchell, R. Howells, E. Wallington, A.J. Fleming and J.E. Gray. 2019. Reduced stomatal density in bread wheat leads to increased water -u se efficiency. *J. Exp. Bot.* 70(18): 4737–4748. https://doi.org/10.1093/jxb/erz248.

Eum, S.M., S.Y. Kim, J.S. Hong, N.S. Roy, S. Choi, J. Paik, S.W. Lee, T.B. Tran, V.H. Do, K.S. Kim and E.S. Seong. 2019. Transcriptome analysis and development of SSR markers of ethnobotanical plant *Sterculia lanceolata*. *Tree Gen. Genomes*. 15: 1–2.

Galla, N.R. and G.R Dubasi. 2010. Chemical and functional characterization of Gum karaya (*Sterculia urens* L.) seed meal. *Food. Hydrocolloid.* 24: 479–485. https://doi.org/10.1016/j.foodhyd.2009.12.003.

Govindaraj, M., M. Vetriventhan and M. Srinivasan. 2015. Importance of genetic diversity assessment in crop plants and its recent advances: An overview of its analytical perspectives. *Genet. Res. Int.* 2015: 431487. https://doi.org/10.1155/2015/431487.

Hermenegildo, P.D.S., L.M.S. Guimarães, J.L. Badel, A.S. Marques, A.C. Alfenas and M.A. Velloso-Ferreira. 2021. Improved evaluation of the genetic variability of Brazilian strains of *Erwinia psidii* with newly developed microsatellite markers. *Plant. Pathol.* 70(4): 922–931. https://doi.org/10.1111/ppa.13343.

Huang, C., C. Shen, T. Wen, B. Gao, D. Zhu, X. Li, M.M. Ahmed, D. Li and Z. Lin. 2018. SSR-based association mapping of fiber quality in upland cotton using an eight-way MAGIC population. *Mol. Gen. Genomics*. 293: 793–805. https://doi.org/10. 1007/s00438-018-1419-4

Huang, Y., B. Niu, Y. Gao, L. Fu and W. Li. 2010. CD-HIT suite: a web server for clustering and comparing biological sequences. *Bioinformatics* 26(5): 680–682. https://doi.org/10.1093/bioinformatics/btq003

Indu, B., S.K. Mohonty, S. Bhat, M.K. Swamy and M. Anuradha. 2019. Genetic diversity and conservation of *Pterocarpus santalinus* L.f. through molecular approaches. In: Pullaiah, T., Balasubramanya, S., Anuradha, M. (eds.) *Red Sanders: Silviculture and Conservation*. Springer, New York, pp. 173–187.

Jaisankar, I., T. Subramani, A. Velmurugan and A.K. Singh. 2017. Genetic diversity revealed among ratstan genotyles from Andaman and Nicobar Islands based on RAPD and ISSR markers. *Forest Res. Int. J.* 1(2): 45–50.

Jaisankar, I., A. Velmurugan and C. Sivaperuman. 2018. Biodiversity conservation: Issues and strategies for the tropical islands. In: Sivaperuman, C., Velmurugan, A., Singh, A.K., Jaisankar, I., (eds.) *Biodiversity and Climate Change Adaptation in Tropical Islands*. Academic Press, London, pp. 525–552.

Jannah, N., M. Pharmawati and U. Uslan. 2022. Genetic diversity of *Sterculia quadrifida* from Kupang based on ISSR profiles, stomatal density, and chlorophyll content. *Biodiversitas*. 23(5): 1–8.

Jiménez, A., H. Mansour, B. Keller and E. Conti. 2014. Low genetic diversity and high levels of inbreeding in the Sinai primrose (*Primula boveana*), a species on the brink of extinction. *Plant Syst. Evol.* 300(5): 1199–1208. https://doi.org/10.1007/s00606–013–0955-y.

Kloch, M.O., A.N. Hawliczek-Strulak and A.N. Sekrecka-Bielak. 2015. Molecular markers in forest management and tree breeding: a review. *For. Wood. Technol.* 92: 193–199.

Lee, BY., H.S. Kim, B.S. Choi, D.J. Hwang, A.J. Choi, J. Han, W.J. Won, A.I. Choi, S.H. Lee, A.S. Om, H.G. Park and J.S. Lee. 2015. RNA-seq based whole transcriptome analysis of the cyclopoid copepod *Paracyclopina nana* focusing on xenobiotics metabolism. *Comp. Biochem. Physiol. Part D Genomics Proteomics.* 15: 12–19. https://doi.org/10.1016/j.cbd. 2015.04.002

Lidder, P., and A. Sonnino. 2012. Biotechnologies for the management of genetic resources for food and agriculture. *Adv. Genet.* 78: 1–167

Luján-Medina, G.A., J. Ventura, A.C.L. Ceniceros, J.A. Ascacio, D.B.V. Valdés and C.N. Aguilar. 2013. Karaya gum: general topics and applications. *Macromol. Indian J.* 9: 111–116.

Mahajan, R., and Gupta, P. 2012. Molecular markers: their use in tree improvement. *J. For. Sci.* 58(3): 137–44.

Maxted, N., 2001. Ex situ, in situ conservation. In: *Encyclopedia of Biodiversity*. Elsevier, USA, New York, pp. 683–695

Nath, S., S. Majumder, M. Ghorai, V. Jain, A.R. Al-Tawaha, S.M. Shekhawat, D.K. Pandey, S. Paul, M.K. Swamy and A. Du. 2023. Assessment of *Boswellia* species genetic diversity using molecular markers. In: *Frankincense–Gum Olibanum*. Apple Academic Press, London, pp. 241–256.

Nikmah, I.A., R. Azrianingsih and D. Wahyudi. 2016. Genetic diversity of Porang populations (*Amorphophallus muelleri* Blume.) in Central Java and West Java based on LEAFY second intron marker. *J. Trop. Life Sci.* 6(1): 23–27. https://doi.org/10.11594/jtls.06.01.05.

Plenk, K., K. Bardy, M. Höhn and M. Kropf. 2019. Long-term survival and successful conservation? Low genetic diversity but no evidence for reduced reproductive success at the north-westernmost range edge of *Poa badensis* (Poaceae) in Central Europe. *Biodiv. Conserv.* 28: 1245–1265. https://doi.org/10.1007/s10531–019–01722-x.

Poerba, Y.S., D. Martanti and F. Ahmad. 2018. Genetic variation of wild *Musa acuminata* Colla from Indonesia based on RAPD and ISSR markers. *Biotropia* 26(2): 1–18. https://doi.org/10.11598/btb.2019.26.2.896.

Prasad, N., N. Thombare, S.C. Sharma and S. Kumar. 2022. Production, processing, properties and applications of karaya (*Sterculia* species) gum. *Ind. Crops Products.* 177: 114467.

Prasad, N., N. Thombare, S.C. Sharma and S. Kumar. 2023. Recent development in the medical and industrial applications of gum karaya: A review. *Polym. Bull.* 80(4): 3425–3447.

Saragih, GS., and S. Siswadi. 2019. Antioxidant activity of plant parts extracts from *Sterculia quadrifida. Asian J. Pharmaceut. Clin. Res.* 12(7): 143–148. https://doi.org/10.22159/ajpcr.2019.v12i7.33261

Savolainen, O., and T. Pyhäjärvi. 2007. Genomic diversity in forest trees. *Curr. Opin. Plant Biol.* 10(2): 162–167.

Spinedi, N., N. Rojas, R. Storb, J. Cabrera, E. Aranda, M. Salierno, M. Svriz and J.M. Scervino. 2019. Exploring the response of *Marchantia polymorpha*: Growth, morphology and chlorophyll content in the presence of anthracene. *Plant Physiol. Biochem.* 135: 570–574. https://doi.org/10.1016/j.plaphy.2018.11.001.

Swamy, M., and M. Anuradha. 2011. Analysis of genetic variability in patchouli cultivars (*Pogostemon cablin* Benth.) by using RAPD markers. *Res Biotechnol.* 2(6): 64–71.

Uslan, U., and M. Pharmawati. 2020. Genetic diversity of *Sterculia quadrifida* in Kupang, Indonesia based on RAPD (Random Amplified Polymorphic DNA) markers. *Biodiversitas.* 21(7): 3407–3414. https://doi.org/10.13057/biodiv/d210766.

Wahyudi, D., U. Uslan and K. Rifliyah. 2020. Genome evaluation of banana cultivars based on morphological character and Inter-Simple Sequence Repeat (ISSR) molecular marker. *Biodiversitas.* 21(7): 2982–2990. https://doi.org/10.13057/biodiv/d210715

10 Trade and Commerce of Gum Karaya

S. Noorunnisa Begum,
K. Ravikumar, and R. Patturaj

10.1 INTRODUCTION

This chapter is about gum karaya obtained from *Sterculia urens*, a species of the family Malvaceae. It is native to India and has been introduced into Burma. A small- to medium-sized tree with a pale-coloured trunk, it is traded in the name of bhutyā in Marathi, kulu, Indian tragacanth, gum karaya, katira, sterculia gum, or kateera gum. The various regional names for the gum are Indian tragacanth (English), Gulu, Kadaya (Hindi), Balika (Sanskrit), Kempudale (Kannada), Theethondi (Malayalam), Kavalam (Tamil), Errapuniki, and Tanuku (Telugu.).

Karaya gum is a natural gum obtained from the stem of the evergreen tree *Sterculia urens* (Figure 10.1). It is primarily used as an emulsifier in the food and beverage and cosmetic industries. It has a high viscosity and gelling properties, which make it an ideal ingredient for emulsions, creams, lotions, and ointments.

Karaya gum, also known as Sterculia gum, is a natural gum obtained from the stem and branches of the karaya tree. The gum has been used in Ayurvedic and Siddha medicine for centuries. It is now being used in many Western pharmaceuticals. The main active ingredient in karaya gum is sterculic acid, which has anti-inflammatory, antibacterial, and antifungal properties. It is also a powerful laxative. Karaya gum is used in many different types of medication, including ointments, creams, suppositories, capsules, tablets, and powders.

Karaya gum has been used for centuries as a stabilizing and thickening agent in foods and cosmetics. Karaya gum is used as a thickener and stabilizer in various food applications, such as sauces, gravies, ice creams, desserts, etc. Karaya gum is composed of galactose, mannose, and glucuronic acid. These components give it the ability to bind to water and form a gel-like substance. This gel-like substance helps stabilize emulsions and suspensions and thicken creams and lotions.

The main uses for karaya gum are in food, cosmetics, and pharmaceuticals; however, it can also be used as an adhesive. In various culinary products, karaya gum is identified by the E416 number. Additionally, it is employed in the treatment of wounds and bedsores, high plasma lipid levels, the glue for dental implants, adhesive plasters, sore throats, and diabetes (Verbeken et al., 2003).

This species is distributed in the Indo-Malesian region. In India, it is found across tropical regions up to 1,000 m altitude (Figure 10.2).

 DOI: 10.1201/9781003438960-10

FIGURE 10.1 *Sterculia urens* – tree.

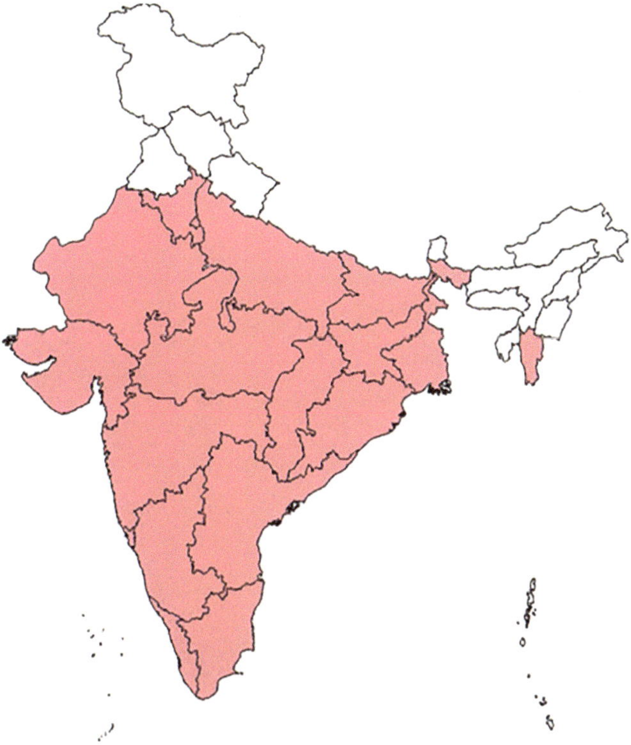

FIGURE 10.2 India map showing the occurrence of *Sterculia urens*.

10.2 DESCRIPTION OF THE RAW DRUG

The raw drug consists of gum which generally resembles gum tragacanth obtained from *Astragalus* spp. The first-grade gum is white, the intermediate is grey or pink, and the lowest grade is dark-brown. Gum, especially when fresh, has an acetous odour (Figure 10.3a and Figure 10.3b).

FIGURE 10.3a *Sterculia urens* fresh gum.

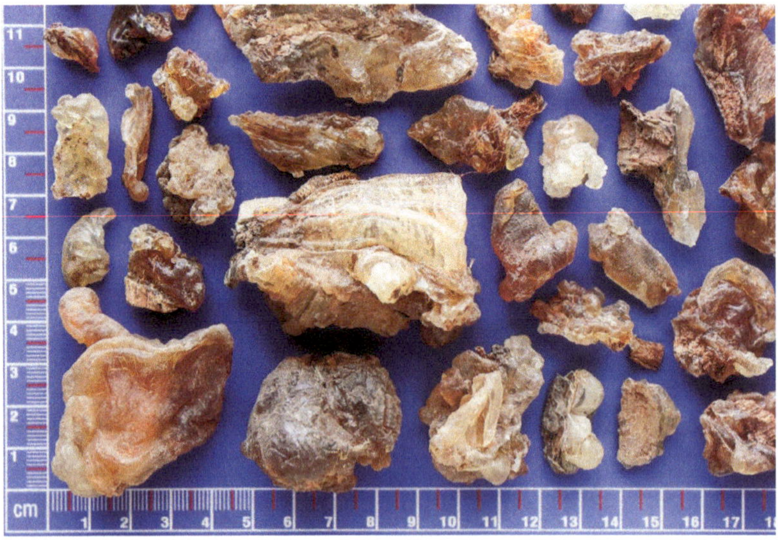

FIGURE 10.3b *Sterculia urens* dried gum as sold in market.

Trade estimated consumption of gum by the domestic herbal manufacturing units has been assessed at a range between 500 and 1,000 MT (dry wt.). Andhra Pradesh reported production of 620 MT (dry wt.), and Madhya Pradesh recorded 110 MT (dry wt.) during 2005–2006, while Orissa reported production of 44 MT (dry wt.) during 2003–2004. During 2004–2005, India exported 832 MT of karaya gum (Indian tragacanth hastab) for a value of 984.2 lakhs (Ravikumar et al., 2018).

10.3 GUM KARAYA AND ITS USAGE IN TRADE

Gum karaya, katira gum, or Indian gum tragacanth obtained from *Sterculia urens* was used as the best substitute for *Astragalus gummifer*, a shrub native to South-west Asia, a true source of gum tragacanth used mainly as a natural emulsifier and thickening agent in drugs and food industries (FRI, 1972; Saha et al., 2017).

Initially, it was sold as an adulterant of true gum tragacanth in the USA, but it has gained unique commercial importance since 1920 (Krishnamurthy, 2010). Now, India is the leading producer and exporter of gum karaya in the world. The other substitutes of gum tragacanth are gum exudates of *Cochlospermum religiosum* and, up to some extent, hog gums from *Bombax ceiba* and *Moringa oleifera* (FRI, 1972).

10.3.1 Differentiating Authentic and Substitute

Gum Karaya

- Colour of the gum karaya is pale yellow or grey to pinkish brown. It has acetous or vinegar odour, acetous or vinegary taste, and it swells in normal water and forms a gel. Molecular weight is 10^6 to 10^7 (MW), and viscosity (cP) is 800 to 1,200 (3% solution).
- The gum exudate of *Sterculia urens* is translucent, pale yellow or grey to pinkish brown in colour, with acetous or vinegary odour and taste (EFSA, 2016).
- This gum can swell more than 60 times of its original volume by absorbing water (Mayes, 2010).
- The viscosity of gum karaya decreases and solubility increases on heating and aging, as a result of the alterations of its polymer conformation (Mayes, 2010).
- Chemically, the exudate of gum karaya is composed of partially acetylated (8–1.4% acetyl groups) complex polysaccharides mixed with calcium and magnesium salts and 1% protein (Mayes, 2010). The chemical structure of this gum consists of a backbone of D-galactose, L-rhamnose, and D-glucuronic acid residues branched with glucuronic acid (Sao, 2012). This gum has high proportion of L-rhamnose (about 15%) and uronic acids (about 40%) in comparison (Mayes, 2010; Sao, 2012).
- The normal pH of 1% gum karaya solution ranges from 4.5 to 4.7 due to the presence of uronic acid residues.

***Astragalus gummifer* Gum**

- Gum obtained from *Astragalus gummifer* is odourless, tasteless, translucent, and off-white to pale yellow in colour (Mayes, 2010; Kulkarni and Shaw, 2016).
- The gum tragacanth swells and forms viscous solutions in both normal water and hot water (Sao, 2012). Its viscosity increases and solubility decreases on heating and aging.
- Tragacanthic acid consists of a backbone of 1,4-linked c-D-galacturonic acid connected with the side chain of 1,3-linked D-xylose, L-fucose, or D -galactose, whereas tragacanthin contains a backbone of 1,6-linked D-galactose units attached with side chains of L-arabinofuranose residues (Mayes, 2010, Emam-Djomeh et al., 2019).
- The gum tragacanth is a complex mixture of 60–70% tragacanthic acid (bassorin), a water-swellable polymer, and tragacanthin (8–10%), a water-soluble arabinogalactan polysaccharide, along with calcium, magnesium, and potassium salts, proteins (<4%), and traces of starch and cellulose.
- pH of 1% gum tragacanth solution varies from 4.5 to 6.0 (Agrawal and Konno, 2009; Emam-Djomeh et al., 2019).

10.3.2 Differentiating Test for Authenticity of the Gum Karaya and Gum Tragacanth

Test for Gum Karaya

1. Boil mucilage with 20% HCl; one gets pink colour after cooling.
2. Take mucilage with sodium hydroxide, and warm. You get brown colour (Chowdhury, 2020).
3. Gum karaya swells in normal water and forms a thick, soft, viscous gel-like colloidal dispersion due to the presence of the acetyl group (Sao, 2012).

Test for Gum Tragacanth

1. Take mucilage and add sodium hydroxide (warm); you get yellow colour.
2. Take mucilage, and add borax and ferric chloride. There is no clear precipitate, but precipitate in clots when alcohol is added to the solution (Parry, 1918).

10.3.3 Chemical Method to Confirm the Identification

Chemical Test

1. To 1 g of powdered karaya gum, add 50 ml of water and mix. A viscous solution is produced, and it is acidic.
2. Add 0.4 g of powdered karaya gum to 10 ml of an ethanol–water mixture (3:2), and mix. The powder is swelling.
3. It readily produces a pink colour with a solution of ruthenium red.

Source: www.pharmacy180.com/article/karaya-gum-105/.

10.4 GUM KARAYA TRADE AND COMMERCE

Traditionally, India has been the largest producer and exporter of karaya gum. The global market for karaya gum increased from US$79.4 million to US$103.3 million between 2016 and 2021, with a CAGR of 5.4%. While the historical outlook for karaya gum sales has been positive, future projections are even more optimistic. This is due to the growing popularity of natural and organic products, as well as the increasing demand for plant-based substances in the food and cosmetics industries. As a result of these trends, karaya gum is expected to see significant growth in both sales and profit over the next few years. This will provide a boost to the economy of countries where it is produced, as well as create new opportunities for trade and investment.

The United States accounted for 32.5% of the global market, with a value of US$35.4 million in 2022. The US market for karaya gum is constantly shaping and changing. The main drivers for this change are the uses of karaya gum. Sales of karaya gum have been increasing steadily in recent years. This is likely due to the growing popularity of natural and organic products. Consumers are becoming more aware of the benefits of using natural ingredients and are willing to pay more for these products.

The India karaya gum market will grow at 3.7% CAGR during the forecast period of 2022 to 2032. The major factor driving the market growth is the growing demand for karaya gum in various applications, such as food and beverage, pharmaceuticals, and cosmetics. Moreover, the rising awareness of the health benefits of karaya gum is also fuelling the market growth in India.

The China karaya gum market will grow at 3.4% CAGR during the forecast period of 2022 to 2032. Karaya gum is derived from the sap of the *Sterculia lychnophora* tree, which is native to China. The gum is used as a thickening agent and stabilizer in a wide range of food and industrial applications. Demand for karaya gum has been growing steadily in China due to its expanding food processing industry. However, domestic production has not been able to keep up with this demand, leading to a growing dependence on imports, primarily from India. The Chinese government has recently announced plans to invest in the domestic production of karaya gum in order to reduce its reliance on imports.

According to the research of Future Market Insights, the market for karaya gum is anticipated to expand at a CAGR of 5.9% over the forecasted period. According to projections, the industry would be worth US$193.2 million by 2032, up from US$108.9 million in 2022. At the end of 2021, the karaya gum market was valued at US$103.3 million. Global market share of the United States, Germany, Japan, and Australia is around 7–8%.

The merger and acquisition of karaya gum suppliers with end-use industries like food and pharmaceuticals are a significant trend for the worldwide karaya gum market

10.5 THREATS TO THE TREE AFFECTING YIELD OF GUM KARAYA

The factors affecting the yield of gum karaya are:

1. Poor fruit set due to self-incompatibility.
2. Pollination happened only by *Apis indica*, and its nectar unavailability causes only 56% pollination efficacy.

3. Absence of wind pollination leaves seeds the only viable mode of propagation.
4. Germination problem: (i) seed germination is poor, freshly harvested (100%) > 10 months, (70%) > 18 months, (60%) > 2 years, and (42%) > 3 years; (ii) hard seed coat faces both physical and physiochemical dormancy.
5. Lack of awareness regarding scientific and advanced gum tapping leads to fatal death (Figure 10.4a, 10.4b, and 10.4c).

FIGURE 10.4a Healthy stem before harvesting.

FIGURE 10.4b Stem after harvesting.

FIGURE 10.4c Close of brutally damaged stem post-harvest.

6. Over collection of seed for consumption.
7. Tree overexploitation for fibre extraction purposes by tribal groups. Young trees reach gum-producing capability at 41.9 cm girth breast height and experience high traditional (brutal) tapping and less gum production (Shukla and Modi, 2020).

Gum tapping is a brutal technique; consequently, due to traditional tapping techniques, insubstantial distribution, illicit gum exploitation, insignificant seed germination, and survival rate, the tree in the Aravalli hills of Rajasthan is enlisted as an endangered plant species (Purohit and Dave, 1996; Reddy and Reddy, 2008).

S. urens gum is employed as a dye thickener for direct-colour printing and textile printing in the textile industry (Patwardhan and Ramachandran, 1945). It is mainly used for cosmetics, lozenges, jellies, emulsions, lotions, sprays, pastes, laxatives, and diarrhoea control in pharmaceutical and medicinal preparations. For the baking and dairy industry, it is used in binding, dressing spread, for healthcare denture fixtures due to these welling and contracting properties in moisture presence. In the paper industry, long-fibred, lightweight papers are produced using it (Dikshith et al., 1984). In general, it is used for lactic products, dressing, whipped cream, and frozen dessert, poorly used in Carnic industry, laxative elaboration, dental adhesive, release matrix of diltiazem, quetiapine fumarate, and delivery of essential oil in patches (Prajapati et al., 2013; López-Franco et al., 2009; Hoefler, 2004; Tamime et al., 2007; Deepthi et al., 2012; Reddy et al., 2012; Kumar et al., 2011; Surendra et al., 2012). It is commercially available from India. The European Union and EFTA permit to use E numbers as codes for food additives. Karaya gum has E number E416 (Anderson, 1988). *S. urens* is widely used for various commercial products due to its unique properties (Shukla and Modi, 2020).

10.6 HISTORICAL OUTLOOK VERSUS FUTURE PROJECTIONS FOR KARAYA GUM MARKET

The global market for karaya gum increased from US$79.4 million **to** US$103.3 million between 2016 and 2021, with a CAGR of 5.4%. While the historical outlook for karaya gum sales has been positive, future projections are even more optimistic. This is due to the growing popularity of natural and organic products, as well as the increasing demand for plant-based substances in the food and cosmetics industries. As a result of these trends, karaya gum is expected to see significant growth in both sales and profit over the next few years. This will provide a boost to the economy of countries where it is produced, as well as create new opportunities for trade and investment. The main uses for karaya gum are in food, cosmetics, and pharmaceuticals; however, it can also be used as an adhesive.

However, the growth of the karaya gum market may be constrained by environmental concerns about the unsustainable harvesting of this tree resin.

During the forecast period, the global karaya gum market is anticipated to grow from US$108.9 million in 2022 to US$193.2 million by 2032, with a healthy CAGR of 5.9%.

10.6.1 Country-Wise Insights

10.6.1.1 How Is the USA Karaya Gum Market Shaping?

The USA accounted for 32.5% of the global market with a value of US$35.4 million in 2022. The USA market for karaya gum is constantly shaping and changing. The main drivers for this change are the uses of karaya gum. Sales of karaya gum have been increasing steadily in recent years. This is likely due to the growing popularity of natural and organic products. Consumers are becoming more aware of the benefits of using natural ingredients and are willing to pay more for these products.

10.6.1.2 How Is India Performing in the Karaya Gum Market?

The India karaya gum market will grow at 3.7% CAGR during the forecast period of 2022 to 2032. The major factor driving the market growth is the growing demand for karaya gum in various applications, such as food and beverage, pharmaceuticals, and cosmetics. Moreover, the rising awareness of the health benefits of karaya gum is also fuelling the market growth in India.

10.6.1.3 What Is China's Position in the Karaya Gum Market?

The China karaya gum market will grow at 3.4% CAGR during the forecast period of 2022 to 2032. Karaya gum is derived from the sap of the *Sterculia lychnophora* tree, which is native to China. The gum is used as a thickening agent and stabilizer in a wide range of food and industrial applications. Demand for karaya gum has been growing steadily in China due to its expanding food processing industry. However, domestic production has not been able to keep up with this demand, leading to a growing dependence on imports, primarily from India. The Chinese government has recently announced plans to invest in the domestic production of karaya gum in order to reduce its reliance on imports.

10.6.2 Category-Wise Insights

10.6.2.1 Why Is Karaya Gum Used in the Food and Beverage and Cosmetic Industries?

Karaya gum is a natural gum obtained from the stem of the tree *Sterculia urens*. It is primarily used as an emulsifier in the food and beverage and cosmetic industries. It has high viscosity and gelling properties, which make it an ideal ingredient for emulsions, creams, lotions, and ointments.

10.6.2.2 How Does Karaya Gum Work as a Stabilizer and Thickening Agent?

Karaya gum has been used for centuries as a stabilizing and thickening agent in foods and cosmetics. Karaya gum is used as a thickener and stabilizer in various food applications, such as sauces, gravies, ice creams, desserts, etc. Karaya gum is composed of galactose, mannose, and glucuronic acid. These components give it the ability to bind to water and form a gel-like substance. This gel-like substance helps stabilize emulsions and suspensions and thicken creams and lotions.

10.6.2.3 What Are the Benefits of Karaya Gum in the Pharmaceuticals Industry?

Karaya gum, also known as Sterculia gum, is a natural gum obtained from the stem and branches of the karaya tree. The gum has been used in Ayurvedic and Siddha medicine for centuries. It is now being used in many Western pharmaceuticals. The main active ingredient in karaya gum is sterculic acid, which has anti-inflammatory, antibacterial, and antifungal properties. It is also a powerful laxative. Karaya gum is used in many different types of medication, including ointments, creams, suppositories, capsules, tablets, and powders (*www.futuremarketinsights.com/reports/karaya-gum-market*).

10.6.3 Karaya Gum Market Report Scope

As per future market insights report, their forecast period of 2022 to 2032 based on historical data of 2017 to 2022 shows that key regions covered North America, Latin America, western Europe, eastern Europe, APEX, Japan, and Middle East and Africa (MEA). Key countries that will be covered are India, USA, Germany, China, United Kingdom, Japan, Denmark, Canada, Thailand, Malaysia, Mexico, Poland, Turkey, Vietnam, Indonesia, Cambodia, France, Italy, Spain, South Korea, Russia, Australia, Brazil, Argentina, and South Africa. The key segments that will be covered are end-use industry, application, and region.

10.6.3.1 Karaya Gum Market by Category

1. By end-use industry, karaya gum market is segmented as:
   - Food and beverages
   - Cosmetics
   - Pharmaceuticals
2. By application, karaya gum market is segmented as:
   - Emulsifier
   - Stabilizer

- Laxative
- Thickening agent
- Aphrodisiac
- Adhesive

3. By region, karaya gum market is segmented as:
   - North America
   - Latin America
   - Western Europe
   - Eastern Europe
   - APEX
   - Japan
   - Middle East and Africa (MEA)

10.6.3.2 Karaya Gum Market Overview (2022 to 2032)

According to the research of Future Market Insights, the market for karaya gum is anticipated to expand at a CAGR of 5.9% over the forecasted period. According to projections, the industry would be worth US$193.2 million by 2032, up from US$108.9 million in 2022. At the end of 2021, the karaya gum market was valued at US$103.3 million.

In various culinary products, karaya gum is identified by the E416 number. Additionally, it is employed in the treatment of wounds and bedsores, high-plasma lipid levels, the glue for dental implants, adhesive plasters, sore throats, and diabetes. The merger and acquisition of karaya gum suppliers with end-use industries like food and pharmaceuticals are a significant trend for the worldwide karaya gum market.

10.7 GUM KARAYA AND HSN CODE 13019016 EXPORTS FROM INDIA

(*Source:* www.volza.com/p/gum-karaya/export/export-from-india/hsn-code-13019016/.)

- As per Volza's India export data, gum karaya and HSN code 13019016 export shipments from India stood at 674, exported by 76 India exporters to 186 buyers.
- India exports most of its gum karaya and HSN code 13019016 to France, Japan, and United Arab Emirates and is the largest exporter of gum karaya and HSN code 13019016 in the World.
- The top 3 exporters of gum karaya and HSN code 13019016 are India with **674** shipments, followed by Mali with 554, and United States at the third spot with 92 shipments.
- Top 1 product categories of gum karaya and HSN code 13019016 exports from India are

1. HSN code 13019016: 13019016

These facts are updated as of 11 December 2023 and are based on Volza's export–import data, India export data of gum karaya and HSN code 13019016, sourced from 70 countries' export–import shipments, with names of buyers, suppliers, top decision-maker's contact information like phone, email, and LinkedIn profiles.

HS code	Product description	Standard qty.	Country of origin	Country of destination
13019016	Karaya >> 180 bags gum karaya pack in new double pp bags with polythene in between; double	9,000	India	France
13019016	karaya >> karaya indian gum karaya indian tragacanth	2,900	India	Pakistan
13019016	karaya >> karaya indian gum karaya indian tragacanth	6,000	India	Pakistan
13019016	karaya >> gum karaya 25 kg	10	India	China
13019016	karaya >> karaya gum powder fcc grade batch no. 01280032019 exp dt.27.03.2021; no. 0 1280032019	25	India	Canada
13019016	karaya >> karaya indian gum karaya indian tragacanth	8,965	India	Pakistan
13019016	karaya >> kathira karaya gum kathira – karaya gum -40- bags detaileded as per document cum packing list; de tails	1	India	Singapore
13019016	karaya >> gum karaya 25 kgs	10	India	China
13019016	karaya >> gum karaya 25 kgs	10	India	China
13019016	karaya >> 360 bags gum karaya pack in new double pp bags with polythene in between; doubl e	18,000	India	France
13019016	karaya >> 180 bags gum karaya no. 2 pack in newdouble gunny bags with polythene inside; new double	9,000	India	France
13019016	karaya >> 180 bags indian gum karaya no. 2 packin new double pp bags with polythene inbetween; pack in; in between	9,000	India	France
13019016	karaya >> karaya indian gum karaya indian tragacanth	5,150	India	Pakistan
13019016	karaya >> karaya indian gum karaya indian tragacanth	9,900	India	Pakistan
13019016	karaya >> karaya indian gum karaya indian tragacanth other de	1,000	India	United Arab Emirates
13019016	karaya >> gum karaya	250	India	Jordan
13019016	karaya >> gum karaya 25 kgs per bag	19	India	China
13019016	karaya >> gum karaya 810	1	India	South Korea
13019016	karaya >> karaya kwt gum karaya-kwt-spl	10,000	India	Turkey

HS code	Product description	Standard qty.	Country of origin	Country of destination
13019016	karaya >> 120 bags gum karaya no 1 grade pack in new double jute bags with polythene inbetween; i n; in between	3,000	India	Japan
13019016	karaya >> karaya adi gum karaya adi	1	India	South Korea
13019016	karaya >> karaya powder gum karaya powder	1	India	South Korea
13019016	karaya >> karaya minor gum karaya gum karaya minor forest produce; pr oduce	500	India	Thailand
13019016	karaya >> gum karaya 15 bags 20 kgs – bag	300	India	United Arab Emirates
13019016	karaya >> karaya minor gum karaya gum karaya minor forest produce; pr oduce	18,000	India	Spain
13019016	karaya >> karaya powder gum karaya powder	1	India	South Korea
13019016	karaya >> karaya powder gum karaya powder	20	India	Taiwan
13019016	karaya >> karaya indian gum karaya indian tragacanth	5,500	India	Afghanistan
13019016	karaya >> karaya minor gum karaya gum karaya minor forest produce; produ ce	50	India	Indonesia
13019016	karaya >> karaya indian gum karaya indian tragacanth	4,000	India	Afghanistan
13019016	karaya >> gum karaya hps gum karaya hps dspg-16827 lot nog 126–210 processing datc02–20 20 expiry date02–2022 produce of india; dspg-1682 7	200	India	Japan
13019016	karaya >> gum karaya hps dspg-00000	200	India	Japan
13019016	karaya >> 200 bags gum karaya no 1 grade pack in new double jute bags with polythene inbetween; i n; in between	5,000	India	Japan
13019016	karaya >> gum karaya pkd in bags	26,000	India	United Arab Emirates
13019016	karaya >> karaya gum indian tragacantgh hastab	200	India	Germany

(Continued)

(Continued)

HS code	Product description	Standard qty.	Country of origin	Country of destination
13019016	karaya >> 200 bags gum karaya no 1 grade pack	5	India	Japan
13019016	karaya >> 180 bags gum karaya no. 2 pack in new double gunny bags with polythene inside; ne w; insi de	9,000	India	France
13019016	karaya >> gum karaya pkd in bags	4,000	India	United Arab Emirates
13019016	karaya >> gum karaya	7,000	India	Afghanistan
13019016	karaya >> 180 bags indian gum karaya no 2 pack	9,000	India	France
13019016	karaya >> 200 bags gum karaya no. 2 pack in new double gunny bags with polythene inside; ne w; insi; de	10,000	India	France
13019016	karaya >> 200 bags gum karaya no. 2 pack in ne	10,000	India	France
13019016	karaya >> gum karaya pkd in bags as per document	13,030.50	India	United Arab Emirates
13019016	karaya >> 180 bags indian gum karaya no 2 packin new double pp bags with polythene inbetween; pack in; in; between	9,000	India	France
13019016	karaya >> 180 bags indian gum karaya no 2 packin new double pp bags with polythene inbetween; pack in; in; between	9,000	India	France
13019016	karaya >> gum karaya 200 bags pack in pp bags other detaileded as per document; oth er	10,000	India	Afghanistan
13019016	karaya >> karaya katira gum karaya katira goond whole	202	India	Canada
13019016	karaya >> natural gum. gum karaya	1,000	India	Brazil
13019016	karaya >> gum karaya hps	400	India	Japan
13019016	karaya >> gum karaya	2	India	Singapore
13019016	gum karaya hps	200	India	Japan
13019016	gum karaya h.p. s. no. 1 grade	10	India	Japan
13019016	gond katira gum karaya	50	India	Germany
13019016	katira goond gum karaya	136	India	Canada
13019016	gum karaya 3kg	3	India	Zambia
13019016	gum karaya 3kg	3	India	Zambia
13019016	gum karaya 3kg	3	India	Zambia

HS code	Product description	Standard qty.	Country of origin	Country of destination
13019016	gum karaya 3kg	3	India	Zambia
13019016	140 bags indian gum karaya no 2 packed	7,000	India	France
13019016	180 bags indian gum karaya no 2 packed	9,000	India	France
13019016	katira goond gum karaya	136	India	Canada
13019016	gum karaya minor forest produce	18,000	India	Spain
13019016	karaya gum gond katira	62.40	India	Australia
13019016	goond katira whole gum karaya	82.50	India	Canada
13019016	goond katira whole gum karaya	82.50	India	Canada
13019016	300 bags indian gum karaya no 2 packe	15,000	India	France
13019016	300 bags indian gum karaya no 2 packe	15,000	India	France
13019016	katira gum karaya gum	4,000	India	Bangladesh
13019016	karaya gum 20 bags 20 kgs/hdpe bag	400	India	United Arab Emirates
13019016	190 bags gum karaya no 2 packed in new	9,500	India	France
13019016	190 bags gum karaya no 2 packed in new	9,500	India	France
13019016	gum karaya packing as per document	2	India	Malaysia
13019016	karaya gum kr150a other details as per	200	India	Poland
13019016	180 bags indian gum karaya no 2 packed	9,000	India	France
13019016	180 bags indian gum karaya no 2 packed	9,000	India	France
13019016	200 bags indian gum karaya no 2 packed	10,000	India	France
13019016	160 bags indian gum karaya no 2 packed	8,000	India	France
13019016	gum karaya	24,000	India	Afghanistan
13019016	other products: gum karaya 2kg	2	India	Zambia
13019016	120 bags indian gum karaya no 2 packe	6,000	India	France
13019016	240 bags indian gum karaya no 2 packed	12,000	India	France
13019016	other products: gum karaya 2kg	2	India	Zambia
13019016	gum karaya – gum karaya minor forest pro	200	India	Indonesia

(Continued)

(Continued)

HS code	Product description	Standard qty.	Country of origin	Country of destination
13019016	120 bags indian gum karaya no 2 packed	6,000	India	France
13019016	240 bags indian gum karaya no 2 packed	12,000	India	France
13019016	karaya gum hps	200	India	Japan
13019016	karaya gum hps	200	India	Japan
13019016	gum karaya packed in double hdpe bag wit	0.50	India	Taiwan
13019016	gum karaya packed in double hdpe bag wit	0.50	India	Taiwan
13019016	180 bags indian gum karaya no 2 packed	9,000	India	France
13019016	160 bags indian gum karaya no 2 packed	8,000	India	France
13019016	200 bags indian gum karaya no 2 packed	10,000	India	France
13019016	180 bags indian gum karaya no 2 packed	9,000	India	France
13019016	gum karaya indian tragacanth	12,200	India	United Arab Emirates
13019016	karaya gum indian tragacanth hastab	200	India	Germany
13019016	karaya gum indian tragacanth hastab	200	India	Germany
13019016	karaya gum indian tragacanth hastab	200	India	Germany
13019016	karaya gum indian tragacanth hastab	200	India	Germany
13019016	karaya gum indian tragacanth hastab	200	India	Germany
13019016	gum karaya – gum karaya minor forest pro	500	India	Thailand

Source: www.volza.com/p/gum-karaya/export/export-from-india/hsn-code-13019016/.

10.8 STATUS OF *STERCULIA URENS* AS NTFP

An asset for the sustenance of tribal based on the 2001 Census report, 8.3%, approximately 8.4 crores of the Indian population are tribes/Adivasis with NWFP as a significant source of their livelihood. Around 5 crore Indians depend on NWFP for livelihood and wage income. To sustain lives, tribal people have been sourcing this plant for diverse purposes, which specifies its significance in their life. NWFP contributes about 50% of the total tribal income in Andhra Pradesh (Reddy and Kumar, 2010). Moreover, the tribal residents are the most vulnerable and disadvantaged communes experiencing extreme economic dismissal. Due to the uncertain income, source of revenue, and lack of awareness, it was lethally used. Corruption is a chief struggle as they earn marginal income from diverse NWFPs mainly because of insufficient knowledge about the commercial value and utilization of these resources along with deficient value addition (Bhattacharya et al., 2003). With the help of NGO and governmental bodies, gradual and steady change is seen in favour of tribal communities. The tribal community experienced gradual social uplift after increased *S. urens* population density, such as (1) livelihoods enhancement and community engagement, (2) increased employability chance, (3) reduced migration of tribes, (4) introduction of social forestry causing financial gain.

10.9 CONCLUSION

Owing to a huge demand and high value of this commercially important gum, karaya trees are being overexploited. To curb the threat to karaya trees in natural habitat, some states in India have banned the commercial production and marketing of this gum. Hence, it is required to establish a balance between gum production and species conservation, which can be done through adopting the scientific tapping methods. This chapter is likely to help research workers, enthusiasts, and end users of natural gums.

REFERENCES

Agrawal, A.A. and K. Konno. 2009. Later A model for understanding mechanisms, ecology, and evolution of plant defense against herbivory. *Ann. Rey. Ecol. Systematics* 40: 311–331.

Anderson, D.M. 1988. The natural gums. *Nutri. Bull.* 13(2): 101–113. https://doi.org/10.1111/j.1467–3010.tb00280.x

Bhattacharya, P., B. Joshi, and S.F. Hayat. 2003. An improved method of tapping gum from kullu *Sterculia urens. Forests Trees Livelihood.* 13(2): 187–196. https://doi.org/10.1080/14728028.2003.9752454

Chowdhury, A.R. 2020. Characterization techniques of natural resins and gums. In: N. Prasad, N. Thombare, R. Singh, P. Sakare and K.K. Sharma (eds), *Training manual on ICAR-NAHEp-CAAST Short training on natural resins and gums: Production, processing value addition and marketing*, pp. 139–146. ICAR-IINRG, Ranchi.

Deepthi, B., S. Mani Kiran, L. Prasanna, and N. Rao. 2012. An investigation of hydrophilic natural gums in the formulation of quetiapine fumarate matrix tablets. *Intern. J. Pharma. Pharmaceut. Sci.* 4(4): 569–574.

Dikshith, T.S.S., R.B. Raizada, R.B. Misra, and K. Srivastava, K. 1984. Toxicological evaluation of karaya gum; acute and subacute oral toxicity in rats. *J. Biosci.* 6(1): 147–153. https://doi.org/10.1007/BF02702866

EFSA, 2016. Re-evaluation of karaya gum (E 416) as a food additive. *European Food Safety Authority J.* 4(12): 4598. https://doi.org/10.2903 I j.efsa.20 I 6.4598

Emam-Djomeh, Z., M. Fathi, and G. Askari. 2019. Gum tragacanth (*Astragalus gummifer* Labillardiere). In: M.A. Razai (ed), *The emerging natural hydrocolloids: Rheology and functions*, pp. 299–326. John Wiley & Sons Ltd, Hoboken.

FRl. 1972. *Indian forest utilization*, Volume-II, The Manager of Publications, Delhi and Forest Research Institute and Colleges, Dehradun.

Hoefler, A.C. 2004. Functions and properties. In: Eagan Press (ed), *Hydrocolloids: [Practical Guides for the Food Industry]*, pp. 27–41. American Association of Cereal Chemist. St. Paul, MN. https://doi.org/10.1094/1891127381.003

Krishnamurthy, T. 2010. *Minor forest products of India (Non-timber forest products)*, BS Publications, Hyderabad.

Kulkarni, V.S. and C. Shaw. 2016. *Essential chemistry for formulators of semisolid and liquid dosages*, Academic Press, New York.

Kumar, A., T. Balakrishna, and J. Rajiv. 2011. Formulation and evaluation of mucoadhesive microcapsules of metformin HCL with gum karaya. *Int. J. Pharm. Pharm. Sci.* 3(3): 150–155.

López-Franco, Y., I. Higuera-Ciapara, F.M. Goycoolea, and W. Wang. 2009. Other exudates: tragancanth, karaya, mesquite gum and larchwood arabinogalactan. In: *Handbook of hydrocolloids*, pp. 495–534. Woodhead Publishing, London. https://doi.org/10.1533/9781845695873.495

Mayes, 1.M. 2010. Gum tragacanth and karaya. In: A. Imeson (ed), *The food stabilizers, thickeners and gelling agents*, pp. 167–179. Blackwell Publishing Ltd., New York.

Parry, E.F. 1918. *Gums and resins: Their occurrence, properties and uses*, Sir Isaac Pitman and Sons, Ltd, London.

Patwardhan, V.C. and S.R. Ramachandran. 1945. Gum *Sterculia urens* as a thickener in textile printing. *J. Indian Chem. Soc. Indus.* 8: 14–17.

Prajapati, V.D., G.K. Jani, N.G. Moradiya, and N.P. Randeria. 2013. Pharmaceutical applications of various natural gums, mucilages and their modified forms. *Carbohyd. Polym.* 92(2): 1685–1699. https://doi.org/10.1016/j.carbpol.2012.11.021

Purohit, S.D. and A. Dave. 1996. Micropropagation of *Sterculia urens* Roxb. an endangered tree species. *Plant Cell Rep.* 15(9): 704–706. https://doi.org/10.1007/BF00231929

Ravikumar, K., S. Noorunnisa Begum, D.K. Ved, J.R. Bhatt, and G.S. Goraya. 2018. *Compendium of traded indian medicinal plants.* Ministry of Environment, Forest and Climate Change (MoEFCC), Government of India, New Delhi and Foundation for Revitalization of Local Health Tradition (FRLHT), Bangalore.

Reddy, M.G. and K.A. Kumar. 2010. *Political economy of tribal development: A case study of Andhra Pradesh.* Centre for Economic and Social Studies, Hyderabad.

Reddy, K.N. and C.S. Reddy. 2008. First red list of medicinal plants of Andhra Pradesh, India-conservation assessment and management planning. *Ethnobotanical Leafl.* 1: 12. https://opensiuc.lib.siu.edu/ebl/vol2008/iss1/12

Reddy, M.M., J.D. Reddy, A. Moin, and H.G. Shivakumar. 2012. Formulation of sustained-release matrix tablets using cross-linked karaya gum. *Trop. J. Pharmaceut. Res.* 11(1): 28–35. https://doi.org/10.4314/tjpr.v11i1.4

Saha, A., S. Tyagi, R.K. Gupta, and Y.K. Tyagi. 2017. Natural gums of plant origin as edible coatings for food industry applications. *Crit. Rev. Biotechnol.* 37(8): 959–973.

Sao, K.P. 2012. *Physico-chemical properties of some Indian plant gums of commercial importance.* ICAR-indian Institute of Natural Gums and Resins, Ranchi.

Shukla, A. and N. Modi. 2020. India's Globally Important *Sterculia urens* Roxb. in crisis: critical insights into conservation, prospects and challenges. *Indian J. Nat. Sci.* 10(61).

Surendra, M., T. Venkateswara Rao, K. Lokeswara Reddy, and A. Ramesh Babu. 2012. Formulation and Evaluation of Mucoadhesive Microcapsules of Aceclofenac for Oral controlled release by Emulsification-gelation Technique. *Res. J. Pharm. Tech.* 5(4): 525–532.

Tamime, A.Y. and R.K. Robinson. 2007. *Tamime and Robinson's yoghurt: Science and technology*. Elsevier, London.

Verbeken, D., S. Dierckx and K. Dewettinck. 2003. Exudate gums: Occurrence, production, and applications. *Appl. Microbiol. Biotechnol.* 63(1): 10–21.

11 Gum Karaya – Threats and Conservation

Lepakshi Md. Bhakshu and K. Venkata Ratnam

11.1 INTRODUCTION

The genus *Sterculia* (family Malvaceae) is represented by 60 species throughout the tropical world and is famous as gum karaya, an alternative to the tragacanth, and often termed as Indian tragacanth. *Sterculia urens* and *S. foetida* are deciduous, native tree species of India standing for major yield per cent of gum karaya. The plant is widely recognized as 'gum karaya' because of its exudates, which are used in the trade. The most ideal environment for this type of tree is tropical, dispersed, muddy rocks at elevations between 1,300 and 2,600 ft. Due to its ability to naturally secrete gum from its stems, *S. urens* is very significant in terms of its ethnobotanical implications and plays a vital role in the life of the tribal people (Mamta et al., 2020; Purohit and Dave, 1996).

This plant is widely used in ethnobotany to treat a wide range of illnesses, including dyspepsia, rheumatoid arthritis, leucorrhoea, amoebic dysentery, and oligospermia. Additionally, market goods have long been accessible in the waste management, food, pharmaceutical, paper, textile, composite fibre, and leather sectors (Venkata Ratnam and Venkata Raju, 2005; Rao et al., 2011; Mamta et al., 2019). The insights of the last chapter deal with the therapeutic characteristics of the *Sterculia* species. On the other hand, not much research has been done on the propagation techniques that serve as the cornerstone of species conservation.

Various species of gum karaya are utilized as additives in culinary, pharmaceutical, cosmetic, paper, textile, composite fibre, waste management, and leather applications. The gum karaya market is expected to reach an economic worth of US$90.1 million by 2025, based on market surveys and research conducted by various organizations (Future Market Insights, 2020; Persistence Market Research, 2020). Several businesses, including Sigma-Aldrich Co. LLC, Gum Technology Corporation, AEP Colloids, Parchem Fine & Specialty chemicals, Andina Ingham Ltd., PPA Inc., Kachabo Gums, DK Exports, and Girijan Co-operative Corporation, are helping to reach the anticipated goal.

Plant establishment in nature is relatively difficult, owing to physical and physiological variables, despite its considerable commercial relevance. This species reproduces only by seeds in the wild. On the other hand, a lengthy time may have passed between dispersal and the beginning of the favourable circumstances necessary for seed germination, which might account for low germination frequency and poor seed viability. Furthermore, because seeds have a high nutritional value, browsing animals in the

DOI: 10.1201/9781003438960-11

forest (colobines, etc.) eat them; as a result, very few seeds remain in the natural habitat to germinate further. Occasionally, even when seeds are given favourable germination circumstances, sheep and goats may graze the young plants, which prevents seedlings from growing quickly enough to regenerate at ground level (Orwa et al., 2009).

Because of the severe limitations imposed by the natural environment, traditional seed propagation is thus not a suitable means of meeting the growing need on an industrial scale (Devi et al., 2012). The key factors contributing to the vulnerable (*Vu*, population reduction: 30 to 49%) and endangered (*En*, population reduction: 50 to 80%) status in some Indian states, including Andhra Pradesh (Vu), Chhattisgarh (Vu), Madhya Pradesh (Vu), and Rajasthan (En), are people's lack of awareness, their susceptibility to non-scientific methods for gum collection from plants, and ongoing overexploitation (Table 11.1) (Nair et al., 1995; Reddy and Reddy, 2008; Kala, 2016). The Indian government is doing several awareness and preservation campaigns to stop deforestation.

Gum tapping is forbidden by the Forest Department in Madhya Pradesh, India, and it is unlawful to process, transport, store, or market gum illegally in Chhattisgarh (with a few exceptions) (Yogi et al., 2015; Mamta et al., 2019, 2020). The developed methodology by Mamta et al. (2020) can be a valuable tool in producing high-quality plantlets in greater quantities to counteract the natural environment's falling *S. urens* abundance.

TABLE 11.1
IUCN categories of species of *Sterculia*

Sl. no.	Name of the *Sterculia* species	IUCN category	Reference/web resource
1.	*Sterculia africana* (Lour.) Fiori	Least concern	https://en.wikipedia.org/wiki/Sterculia_africana; www.iucnredlist.org/species/62754/146209124
2.	*Sterculia alexandria* Harv.	Rare and endemic to South Africa	http://redlist.sanbi.org/species.php?species=1851-2
3.	*Sterculia apetala* (Jacq.) H. Karst.	Least concern	www.iucnredlist.org/species/61787794/149016023
4.	*Sterculia balanghas* L.	No mentions about the status	–
5.	*Sterculia cinerea* A.Rich.	Near threatened	www.iucnredlist.org/species/36469/10001379
6.	*Sterculia cochinchinensis* Pierre	No mentions about the status	–
7.	*Sterculia foetida* L.	Not evaluated (NE)	https://ntbg.org/database/plants/detail/sterculia-foetida
8.	*Sterculia guttata* Roxb. ex DC.	Not evaluated (NE)	Sankara Rao et al., 2016
9.	*Sterculia konchurangensis* Kieu et al.	New species; data deficient	Kieu et al., 2023
10.	*Sterculia murex* Hemsl.	Least concern	Raimondo et al., 2009

(Continued)

TABLE 11.1 (*Continued*)
IUCN categories of species of *Sterculia*

Sl. no.	Name of the *Sterculia* species	IUCN category	Reference/web resource
11.	*Sterculia parviflora* Roxb.	Low risk or least concern	www.iucnredlist.org/species/33949/9824139
12.	*Sterculia rogersii* N.E.Br.	Least concern	Raimondo et al., 2009
13.	*Sterculia setigera* Delile	Least concern	www.gbif.org/species/5406681
14.	*Sterculia tragacantha* Lindl.	Least concern	*https://commons.wikimedia.org/wiki/Category:Sterculia_tragacantha*
15.	*Sterculia urens* L.	Vulnerable	Reddy and Reddy, 2008

11.2 CHALLENGES FOR THE CONSERVATION OF *STERCULIA* SPECIES

Since *S. urens* was reported to have recalcitrant seeds which germinate 100% during fresh conditions, the capacity of germination of seeds might be subsequently decreased to 20%, which became a challenge for long storage for propagation (Mamta et al., 2019). Even the techniques of vegetative propagations, namely, grafting, cuttings, budding, and air layering, are also limited and unsuccessful (Mamta et al., 2019), and reports are very scant on the vegetative method of proliferation of *Sterculia* species.

11.2.1 Key Challenges during Reforestation of *S. urens*

When in forest, *S. urens* confronts multiple problems contributing to the critical challenges during its conservation. Sunnichan et al. (2004) stated the major hindrance for the germination to be due to its poor fruit set and self-incompatibility. Besides, pollination is only caused by *Apis indica*, and its nectar unavailability leads to only 56% pollination efficacy (Sunnichan et al., 2004). The major challenge is, seed germination rate in stored seed is poor, freshly harvested (100%) > 10 months, (70%) > 18 months, (60%) > 2 years, and (42%) > 3 years. Hard seed coat also experiences physiochemical and physical dormancy (Devi et al., 2012). Furthermore, ignorance of sophisticated and scientific gum tapping results in fatalities. Over-harvesting of seeds for human consumption and overexploitation of trees for fibre extraction by indigenous people (Sukumaran and Raj, 2010) led to its population decline. According to Mishra et al. (2012), young trees exhibit reduced gum production and increased traditional (unscientific) tapping when they reach 41.9 cm girth breast height.

11.3 SUSTAINABLE CONSERVATION TECHNIQUES: APPROACH FOR A BETTER FUTURE

Germination. Seeds that are picked fresh have a 100% germination rate, which may be maintained for up to ten months before declining to 70%. If the normal moisture level needed for seed germination is maintained, the best germination ability can last for an extended amount of time. In addition, seeds must be stored in a polythene bag at the ideal temperature of 0–4°C to preserve the moisture content necessary for germination (Sunnichan et al., 1998, 2004; Devi et al., 2012; Damle, 2014).

Dormancy. This plant has both chemical and physical dormancy in addition to the difficulty with germination. Acid and mechanical scarification can be used to break physical dormancy, and phyto hormone GA therapy has been shown to be helpful for chemical treatment (Devi et al., 2012).

Scientific tapping. In order to preserve the mature tapped karaya trees and increase output, this method is sustainable. A massive study conducted by Bhatnagar et al., (2012) revealed that about 30% of trees were damaged severely due to pollarding, lopping, and disease, suggesting that scientific gum collection will reduce damage to gum karaya trees. Hence, non-destructive, scientific gum collection methods are recommended to conserve the existing stands for the future generations. Hence, training of gum collectors is essential besides massive plantations.

Nursery technique. Pandey and Khatoon (1999) showed that when seeds are placed 2 cm deep and vertically aligned, they enhance the germination rate to 80%, seed vigour, and dry biomass of seedlings. Devi et al. (2012) provided the protocols for retaining seed viability through proper storage conditions and chemical treatments for its physical and physiological dormancy. These studies might be applied for raising plantlets in silviculture.

11.4 *STERCULIA URENS* ROXB.: A BOON FOR THE TRIBAL ECONOMY

1. According to the 2001 Census report, 8.3% of India's population, or 8.4 crores, are Adivasis and tribes whose primary source of income are the Non Wood Forest Produce (NWFP). Approximately 5 crore Indians rely on NWFP for their ways and means of subsistence. Tribal people have been using this plant for a variety of purposes to support their lifestyles, indicating the plant's importance to them. Approximately half of Andhra Pradesh's total tribal income comes from NWFP (Reddy and Kumar, 2010). In addition, the tribe's citizens are the most marginalized and at risk, facing severe economic discrimination. It was utilized lethally because of the erratic income, lack of understanding, and source of revenue. One of their key challenges is corruption, since they receive meagre revenue from various NWFPs mostly because to inadequate value addition and ignorance of the economic worth and application of these resources (Bhattacharya et al., 2003). Change is occurring gradually and steadily in favour of indigenous groups with the assistance of NGOs and governmental organizations. After *S. urens* population density grew, the tribal group gradually experienced social uplift, including:

   a. improved livelihoods and community participation;
   b. greater employability
   c. decreased tribe migration and
   d. enhanced financial gain due to the introduction of social forestry.

11.5 MICROPROPAGATION TECHNIQUES DEVELOPED FOR CONSERVATION

An in vitro procedure for *S. urens* (gum karaya tree) mass multiplication (micropropagation) has been developed using nodal segments from cotyledons (Purohit and Dave, 1996; Hussain et al., 2007), and somatic embryogenesis (Sunnichan et al., 1998) has been reported. Hussain et al. (2008) developed a micropropagation method using intact seedlings for in vitro shoot proliferation for the fast production of *S. urens*. Devi et al. (2014) suggested the 'synthetic seed technology' provide an alternative for micropropagation which might be useful in germplasm conservation of *S. urens*. The vegetative as well as in vitro propagation protocols for the production of plantlets of *Sterculia* species were elaborately discussed in the earlier chapters. These techniques were scientifically introduced by governmental and non-governmental agencies besides forest authorities.

11.6 CASE STUDY ON *S. FOETIDA*

Kepuh (*S. foetida*), a versatile plant, is presently in danger of becoming extinct. The numerous illegal logging operations, the conversion of the land, and the dormancy of their seeds pose a threat to the conservation of kepuh in the wild. Because of these factors, conservation efforts related to the geographical distribution and association of the kepuh population need to be made known in advance. Because it serves as the basis for addressing associated policy choices and actions, this information is essential. Consequently, research was done by Metananda et al. (2015) to ascertain the kepuh population, its geographic range, and its connections to other species. For this study, surveys and a review of the literature were conducted in 12 districts of Malaysia's Sumbawa Regency. Researchers found 169 kepuh individuals in 12 districts (65 seedlings, 5 saplings, 14 poles, and 85 trees) utilizing the single quadrat technique and excursions. The dispersion of kepuh indicated that it was not spread equally these days. Since kepuh is a plant that is used to manufacture food and drink, its distribution patterns in both the socio-ecological and socio-economic domains were concentrated in the same places. On the other side of the sociobiological clumped distribution, kepuh and other plants show signs of social interaction or affiliation. The association patterns that are shown are frequently positive in character. However, these links were formed at a rather low level, according to the Jaccard index, which assesses connections at the association level (average value of 0.38). The foundation for the sustainable management of kepuh in nature would be the population-supplied information, distribution, and distribution patterns and relationships of kepuh.

The studies of Metananda et al. (2015), an inventory of single plots measuring 1 ha each in three districts, namely, Kec. Moyo Utara, Kec. Lenangguar, and Kec. Empang, found as many as 80 kepuh individuals at the seedling, sapling, pole, and tree levels, or density around ± 27 individuals/ha (Figure 11.1). Depending on the diameter class of the kepuh population (poles and trees), in Kab. Sumbawa, it reaches a diameter of > 100 cm. Most of the blisters were found with a diameter of 20–29 cm, and the diameter above it for multiples of 10 cm continued to decrease. On the basis of growth rate, it was found that the rate of regeneration of the blisters did not run

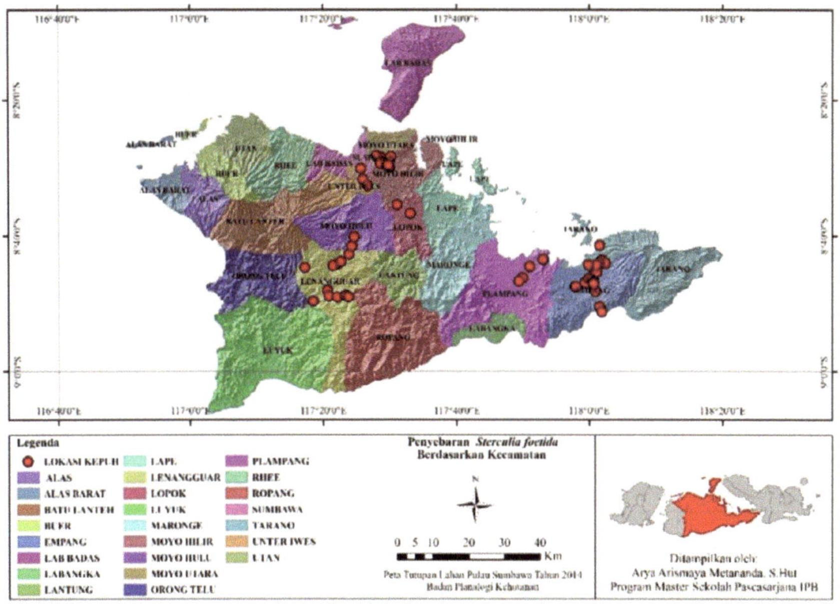

FIGURE 11.1 A satellite topography deciphering the distribution of *S. foetida* in the kepuh, map based on subdistricts in Kab. Sumbawa.

Source: Metananda et al. (2015).

normally. The high density at the seedling level, namely, 61 individuals/ha, was not followed by growth at the sapling level.

The pace at which kepuh regenerates is not working properly. Development issues arise not only in the sapling-to-pole transition but also in the seed-to-new-tiller development. The first issue is that not all seeds fall and have the ability to grow into seedlings; instead, just a portion of the seeds fall and are counted as possible new tillers. The second issue is the propensity to disrupt the blisters' pace of regeneration during sapling development. Another issue is that illnesses and bugs frequently target kepuh (Metananda et al., 2015).

Several factors contribute to the inability of the kepuh seeds to develop, namely, the disturbance of wildlife, livestock, and pest attacks. Kepuh seeds, while still on the tree, are often eaten by squirrels and other types of rodents, even after they fall to the ground; also, the kepuh seeds are sometimes eaten by pigs. Not to mention people's livestock, such as goats, buffaloes, cows, and horses, which often take shelter under kepuh trees, causing kepuh seeds that fall around the brood stock to not be able to live because they are constantly trampled on or disturbed by the livestock's activities. Findings in the field suggest that in dry areas, the fallen kepuh seeds cannot grow. The kepuh seeds might require treatment for breaking of dormancy. Kepuh seeds that have at least three layers of coat slow down the seeding process. According to Sumantri and Supriatna (2010) and Zanzibar (2011), it is suspected that kepuh seeds have skin (seed coat) dormancy, so pre-treatment is suggested during its germination.

11.7 *S. URENS* RESTORATION AND CONSERVATION STRATEGIES: FUTURE SCOPE

Launching innovative technologies is necessary to counteract the declining *S. urens* population (Shukla and Modi, 2020). Here are some strategies for a bright future for this species:

1. Compared to regular seeds, artificial seed technology may be a viable means of obtaining high-grade, healthy somatic embryos that can produce high-frequency karaya plants.
2. One more method for the propagation and growth of high-yielding trees might be the micrografting technique.
3. One plant tissue culture method that might be extremely helpful for the preservation of endangered species is embryo culture.
4. Micropropagation of an elite tree and genetic diversity analysis utilizing molecular markers are important decisions for the *S. urens* tree enhancement program.
5. Because of its potential for social forestry, the Ministry of Forestry must strongly advise farmers and tribal members to plantations of *Sterculia* spp. for agroforestry. Mishra (2012) documented inadequate regeneration in the current stands in addition to non-sustainable gum harvesting.
6. Training programmes to selected licenced gum collectors on scientific blazing and tapping, drying, and storage of gum appropriately until its transport are recommended by Bhatnagar et al. (2012) while studying in *Sterculia* stands in various districts of Madhya Pradesh of India. Bhatnagar and Urmaila (2014) also noticed that inflicting deep and wide blazing also affects kullu tree health, leading to severe tree damage. Daily freshening of blazes also harms the gum-secreting canals in the non-collection seasons. These seriously affect the quality of gum due to contamination, resulting in poor yield.
7. Bhatnagar and Urmaila (2014) recommended identifying the hot spots of kullu trees and noting high-quality germplasm centres for future ideal plants, in addition to preventing harvesting below 90 cm stems (girth diameter) and monitoring gum collection activities.

11.8 THREATS AND MANAGEMENT WITH REFERENCE TO *S. FOETIDA*

The growth of seedlings is fairly rapid in earlier stages, but later, it slows down. *S. foetida* requires light and occupies lots of space for development.

- Germplasm management studies revealed that the seed storage behaviour is orthodox – no problems on long-term storage.
- In India, *S. foetida* is severely affected by the larvae of *Sylepta balteata* (pest), leading to high mortality in the nurseries. The feeding, the larva, produces silky threads, rolls and spins up the leaf, and damages the tissues

between the veins. The larva moves to other portions of the leaf for feeding. The incidence of the pest varies from 70 to 80% (Orwa et al., 2009).

- Decay in damaged stands (*S. foetida*) due to fungal infection with brown root rot species (*Phellinus noxius*) might be another threat to the health of trees, which may affect the population of trees (Lin et al., 2013).

Rai et al. (2020) studied the germination of *S. villosa* from Meerut and Bulandshahr districts (India) with a defined mixture of soil and reported seed germination of 80% in three weeks. The status of seedlings grown as saplings has application in the improvement of *Sterculia* tree nursery (Rai et al., 2020).

11.9 ACTION PLAN

The implication is for managing the kepuh population's future based on low population rate considerations. The distribution tends to be sporadic, and grouping at a particular location is necessary for cultivation (nursery), as an intensive step, particularly in increasing the chances of survival (population kepuh) from seedling to sapling growth rate and so on. In addition, demanding increased supervision at locations where kepuh were found or their habitat preference based on the characteristics of the current habitat was found by related parties, especially from the action threatening illegal logging, for the sake of kepuh sustainability in the future. Moreover, the conservation of gum karaya trees throughout the globe requires suitable action plans for improving local populations in the forested as well as buffer zones, open places, and non-cultivated lands, avenue plantations. The training programs to the tribal people, unemployed youth, or farmers for gum collection and management of non-timber forest produce in toto.

11.10 SUCCESS STORY IN RELATION TO *S. URENS*

The Forest Department (government of Andhra Pradesh), Kovel Foundation (Visakhapatnam) Girijana Cooperative Corporation (GCC), Vana Samrakshana Samithi (VSS), and village-level forest conservation committees made an incredible contribution in conserving the trees and in community mobilization, providing training on building skills on forest products to gum pikers on scientific tapping and methods of enhancing gum yield through application of ethephon. Training on the collection, grading of gum, and providing affordable price are key contributions of Kovel Foundation besides massive gum karaya tree plantations (Srinivas and Umesh Rao, 2007–2008; *https://ntfp.org/2009/10/kovel-foundation-a-journey-towards-sustainable-tribal-livelihoods/*).

11.11 CONCLUSION

The intensified utilization of gum karaya in preparing diversified products, the resource tree stands, leads to overexploitation, one of the major causes of its threat of existence, since the extraction of gum has become unscientific by the local vendors. If this phenomenon continues, the species may become extinct locally. Moreover, the seeds are the only method for its natural regeneration, and the rate of germination also declines during lapse. Hence alarming for its immediate need to conserve by propagating vigorously. Numerous techniques, including mass and clonal/vegetative propagation of better tree genotypes, molecular breeding, artificial seed technology, micrografting, genetic transformation, and genetic diversity analysis for in vitro propagation, may be used for conservation. The in vitro cultivation method is one of the strategies that is most effectively attributed with conserving germplasm, especially when it comes to rare or endangered species. As the karaya gum tree is vital to the tribal economy and has a high economic value, it is also considered an endangered species, and as a result, there is an urgent need to establish intensive conservation management efforts to prevent the tree's extinction. It may become extremely endangered if it is not conserved.

According to the International Union for Conservation of Nature's (IUCN) Red List, *S. foetida* is classified as least concern (LC), meaning, that it is stable, safe, and not in danger of going extinct. This is counterbalanced by the fact that these plants are found practically everywhere in the planet. However, in addition to physiological considerations, it poses local hazards (web source: *https://uptpth.dishut.jatimprov.go.id/kepuh-sterculia-foetida/*). Moreover, the many species of *Sterculia* are not evaluated, except *S. foetida* and *S. urens*, attracting urgent need for its population status as well as development of localized action plans for its conservation.

REFERENCES

Bhatnagar, P., M. Goswami, and S. Jain. 2012. Conservation and Harvesting of *Sterculia urens* (Kullu) Gum in Madhya Pradesh. *Vaniki Sandesh* 3: 1–6.

Bhatnagar, P. and R. Urmaila. 2014. Status of *Sterculia urens* Roxb. trees in Madhya Pradesh, India. *Journal of Non-Timber Forest Products* 21(4): 207–210.

Bhattacharya, P., B. Joshi, and S.F. Hayat. 2003. An improved method of tapping gum from kullu *Sterculia urens*. *Forests, Trees and Livelihoods* 13(2): 187–196. https://doi.org/10.1080/14728028.2003.9752454

Damle, V. 2014. Studies on germination behaviour and propagation in karaya Gum (*Sterculia urens* Roxb.) (Doctoral dissertation, Indira Gandhi Krishi Vishwavidyalya, Raipur. http://krishikosh.egranth.ac.in/handle/1/76986)

Devi, P.S., B. Satyanarayana, A. Arundhati, and T.R. Rao. 2012. Effect of storage temperature and dormancy-breaking treatments on seed germination, moisture content and seed vigor in gum karaya (*Sterculia urens* Roxb.), *Forest Science and Technology* 8(1): 11–15. https://doi.org/10.1080/21580103.2012.658235

Devi, P.S., B. Satyanarayana, A. Arundhati, A., and T.R. Rao. 2014. Plant regeneration from encapsulated nodal segments of *Sterculia urens* Roxb., an endangered gum-yielding tree. *Journal of Crop Science and Biotechnology* 17: 21–25. https://doi.org/10.1007/s12892-013-0097-3

Future Market Insights. 2020. *Karaya Gum market research: Global Industry analysis & Forecast from 2022 to 2030*. dataintelo.com.

Hussain, T.M.D., T. Chandraousekhar, and G. Rama Gopal. 2007. High frequency shoot regeneration of *Sterculia urens* Roxb. an endangered tree species through cotyledonary node cultures. *African Journal of Biotechnology* 6(14): 1643–1649.

Hussain, T.M.D., T. Chandrasekhar, and G. Rama Gopal. 2008. Micropropagation of *Sterculia urens* Roxb., an endangered tree species from intact seedlings. *African Journal of Biotechnology* 7(2): 95–101.

Kala, C.P. 2016. Important Gum Yielding Species *Anogeissus latifolia* (Roxb.) Bedd., *Boswellia serrata* Roxb. and *Sterculia urens* Roxb.: Ethnobotany, population density and management. *Applied Ecology and Environmental Sciences* 4: 61–65. https://doi.org/10.12691/aees-4-3-2

Kieu, C.N., D.B. Tran, N.H. Le, T.H. Duong, T.H. Bui, T.T. Nguyen, H.Q. Bui, and T.B. Tran. 2023. A new species of *Sterculia* (Malvaceae) from Vietnam. *PhytoKeys* 227: 1–8. https://doi.org/10.3897/phytokeys.227.101754

Lin, C.J., C.H. Chung, M.L. Wu, and C.L. Cho. 2013. Detection of *Phellinus noxius* decay in *Sterculia foetida* tree. *Journal of Tropical Forest Science* 25(4): 487–496.

Mamta, D., L. Sharma, S, Singh, and M.M. Sharma. 2020. Ex situ conservation using in vitro methods of an endangered plant *Sterculia urens* Roxb.: A high volume trade plant for Gum Karaya, *Industrial Crops and Products* 158, 13015. https://doi.org/10.1016/j.indcrop.2020.113015.

Mamta, D., A. Singh and M.M. Sharma. 2019. A review on *Sterculia urens* Roxb.: A boon to the livelihood for tribal people and industry. *Industrial Crops and Products* 130: 341–351. https://doi.org/10.1016/j.indcrop.2018.12.065.

Metananda, A.A., E.A. Zuhud, and A. Hikmat. 2015. Population, distribution of Kepuh (*Sterculia foetida* L.) and its associated in Sumbawa Regency, West Nusa Tenggara. *Media Konservasi* 20(3): 277–287. https://doi.org/10.29244/medkon.20.3.%p

Mishra M. 2012. Plant regeneration and harvesting of Kullu in India: Case of Khatpara, FPC, Sehore, Madhya Pradesh. Book of Abstracts: National Workshop on "Sustainable Management of Natural Gum and Oleoresins".

Mishra, S., N. Behera, and T. Paramanik. 2012. Comparative assessment of gum yielding capacities of *Boswellia serrata* Roxb. and *Sterculia urens* Roxb. in relation to their girth sizes eds. In *International Conference on Anthropogenic Impact on Environment & Conservation Strategy*, Ranchi, pp. 327–330.

Nair, M.N.B., K.R. Shivanna, and H.Y. Mohan Ram. 1995. Ethephon enhances karaya gum yield and wound healing response: a preliminary report. *Current Science*. 69(10): 809–810.

Orwa, C., A. Mutua, R. Kindt, R. Jamnadass, and A. Simons. 2009. *Sterculia foetida* (Page 1 of 5), Agroforest tree database: A tree reference and selection guide version 4.0. www.worldagroforestry.org/af/treedb/; https://apps.worldagroforestry.org/treedb2/AFTPDFS/Sterculia_foetida.PDF.

Pandey, A.K. and S. Khatoon. 1999. Effect of orientation of seed placement and depth of sowing on seedling emergence in *Sterculia urens* Roxb. *Indian Forester* 125(7): 720–724.

Persistence Market Research, 2020. Global karaya gum market to reach US$ 90.1 Mn by 2025. *Persistence Market Research*. 16 January 2018, 06:00 ET. https://rb.gy/v0vdl

Purohit, S.D. and A. Dave. 1996. Micropropagation of *Sterculia urens* Roxb.—An endangered tree species. *Plant Cell Reports* 15(9): 704–706. https://doi.org/10.1007/BF00231929

Rai, Y., P. Kumar, S. Rajput, and P. Paliwa. 2020. Growth and development of rare tree species *Sterculia villosa* Linn. in districts, Meerut and Bulandshahr, U.P., India. *Asian Journal of Research in Agriculture and Forestry* 6(4): 1–8. https://doi.org/10.9734/AJRAF/2020/v6i430109.

Raimondo, D., L. von Staden, W. Foden, J.E. Victor, N.A. Helme, R.C. Turner, D.A. Kamundi, and P.A. Manyama. 2009. Red list of South African plants. *Strelitzia 25*. South African National Biodiversity Institute, Pretoria.

Rao, J.K., T.V.V. Seetharami Reddy, and O.A. Kumar. 2011. Ethnobotany of stem bark of certain plants of Visakhapatnam district, Andhra Pradesh. *Current Botany* 2(5): 1–6.

Reddy, K.N. and C.S. Reddy. 2008. First Red List of Medicinal Plants of Andhra Pradesh, India—Conservation Assessment and Management Planning. *Ethnobotanical Leaflets* 12: 103–107.

Reddy, M.G. and K.A. Kumar. 2010. *Political Economy of Tribal Development: A Case Study of Andhra Pradesh*. Centre for Economic and Social Studies, Visakhapatnam.

Sankara Rao, K., R. Arun Singh, D. Kumar, K. Raja Kullayiswamy and P. Navendu. 2016. *Digital Flora of Eastern Ghats*. http://easternghats.ces.iisc.ernet.in/plants.php?name=Sterculia *guttata*.

Shukla, A. and N. Modi. 2020. India's globally important *Sterculia urens* Roxb. in crisis: Critical insights into conservation, prospects and challenges. *Indian Journal of Natural Sciences* 10(61): 27; 173–182.

Srinivas, S. and A. Umesh Rao. 2007–2008. Annual Report of Kovel Foundation (2007–2008), Visakhapatnam, pp. 37–55.

Sukumaran, S. and A.D.S. Raj. 2010. Medicinal plants of sacred groves in Kanyakumari district Southern Western Ghats. *Indian Journal of Traditional Knowledge* 9(2): 294–299.

Sumantri, O. and N. Supriatna. 2010. *Informasi singkat benih Sterculia foetida L.* BPTH Jawa dan Madura, Sumedang, ID.

Sunnichan, V.G., H.M. Ram, and K.R. Shivanna. 2004. Floral sexuality and breeding system in gum karaya tree, *Sterculia urens*. *Plant Systematics and Evolution* 244(3–4): 201–218. https://doi.org/10.1007/s00606-003-0095-x

Sunnichan, V.G., K.R. Shivanna, and H.Y. Mohan Ram. 1998. Micropropagation of gum karaya (*Sterculia urens*) by adventitious shoot formation and somatic embryogenesis. *Plant Cell Reports* 17: 951–956.

Venkata Ratnam, K. and R.R. Venkata Raju. 2005. Folk medicine used for common women ailments by Adivasis in Eastern Ghats of Andhra Pradesh. *Indian Journal of Traditional Knowledge* 4(3): 267–270.

Yogi, R.K., A. Bhattacharya, A.K. Jaiswal, and K. Alok. 2015. *Lac, Plant Resins and Gums Statistics 2014: At a Glance*. ICAR-Indian Institute of Natural Resins and Gums, Namkum, Ranchi. https://doi.org/10.13140/RG.2.1.3061.8964

Zanzibar, M. 2011. *Atlas Benih Tanaman Hutan Indonesia Jilid II*. Balai Penelitian Teknologi Perbenihan Tanaman Hutan, Bogor, ID.

12 Gum Karaya and People's Livelihood

R. Patturaj, S. Noorunnisa Begum, and K. Ravikumar

12.1 INTRODUCTION

Gum exudates are amongst the oldest natural gums. The major exudate gum – gum karaya – possesses a unique range of functionalities. It has been an important item for usage in the food, pharmaceutical, adhesive, paper, textile, and other industries (Setia et al., 2010).

Gum karaya is one of the most important forest products of India. It has been used in 100 years for its commercial value, through the largest producer and exporter of gum karaya (Sahu et al., 2019). Andhra Pradesh, Bihar, Chhattisgarh, Gujarat, Karnataka, Maharashtra, Orissa, and Rajasthan states are producing the gum. It has been collected by making incisions in the cork of the tree and collecting the emission of gum. The collection of exudates is then dried and processed into several forms. It has numerous properties and application in industries such as food, pharmaceuticals, cosmetics, and textiles. Gum karaya is a complex mixture of polysaccharides, including galactose, rhamnose, and glucuronic acid. It forms a viscous, gel-like substance when mixed with water. It has excellent water-holding capacity and viscosity-enhancing properties, making it useful as a thickening and stabilizing agent in various food and beverage products (Raj et al., 2021). It can be used in applications such as sauces, dressings, dairy products, and confectionery. It has emulsifying properties, which means it helps stabilize and create emulsions by dispersing oil droplets, and vice versa. It is commonly used in salad dressings, margarines, and other food products that require emulsion stability.

12.2 ETHNO USES

Gum karaya is used in herbal preparations and remedies. Tribal communities have a deep understanding of their local flora and often utilize different parts of plants for various purposes. *S. urens*, commonly known as gum karaya, is a tree species found in India. There are some traditional uses of its different parts. The bark has been used in traditional medicine. Tribal peoples traditionally prepare decoctions from the bark and use them for treating various ailments, including gastrointestinal disorders, respiratory infections, and skin diseases. Mixture of bark with pepper is a decoction given orally to treat leucorrhoea (Ratnam and Raju, 2005), cracks of feet by rubbing the bark, decoction of bark to be taken once a day to clear the uterus used

DOI: 10.1201/9781003438960-12

by Korku tribe (Jagtap et al., 2006), labour pain (Panduranga et al., 2011) and orally to maintain menstrual cycle (Jain et al., 2005), rheumatoid arthritis (Rao et al., 2016), wound and throat infections (Oak et al., 2015). Barks are used for rope-making and in coarse cloth (Kala, 2011, 2016).

Gum is extracted from the cork of trunks and branches. It has multiple traditional uses. It is used as a thickening agent in food preparations. Gum exudates are used in foodstuffs as emulsifiers, stabilizers, and thickness (Oak et al., 2015) and to treat blisters, blood dysentery, joint pain, stomach disorder, and throat infection, and in tonics, jams, and confectionary (Luján-Medina et al., 2013). A mixture of the gum with $CaCO_3$ is used as an antidote for snakebite and also in leucoderma and peptic ulcer, menstrual disorders (Jain et al., 2005), and other gynaecological uses (Kala, 2011). A mixture of resin and wheat flour/jaggery is given as nutritious food for abdominal disorders (Dhiman et al., 2019). Additionally, it is utilized in traditional medicine for its laxative properties and as an adhesive in arts and crafts. It can be mixed with water to create a natural adhesive for paper, cardboard, and other materials. It is commonly used in bookbinding, college work, and other artistic applications.

Leaves are used in some tribal communities for their medicinal properties. They are often ground into a paste and applied externally to treat skin infections, burns, wounds, and throat infection (Oak et al., 2015). The leaves are also used in traditional remedies for respiratory ailments and are sometimes consumed as an herbal tea, for wound fractures and cracked skin in leukaemia by the tribals in the Kolli hills (Vaidyanathan et al., 2013). Leaves are use as fodder for livestock of tribal communities (Omkar et al., 2012).

Wood is used by tribals for various purposes. It is a durable and termite-resistant wood, making it suitable for construction, crafting tools, and making furniture. The wood is also used for fuel and charcoal productions. Branches are used as cordage, facilitate child delivery, and are used in childbirth (Kala, 2016).

12.3 ECONOMIC USES

12.3.1 Food Industry

Gum karaya is commonly used as a food additive and stabilizer. It is often found in products such as ice cream, yogurt, confectionery, and baked goods. Gum karaya is used as a stabilizer and thickener. It can improve texture, provide stability, and prevent crystallization in food products. It has excellent thickening and stabilizing properties. It can be used in various food products, including in sauces, dressings, gravies, and dairy products, to provide a smooth and creamy texture. It helps prevent phase separation and maintains the consistency of the food (Anderson, 1989).

Gum karaya has emulsifying properties, which means it helps mix and stabilize immiscible substances like oil and water (Shekarforoush, 2016). It can be used in salad dressings, mayonnaise, and other emulsion-based products to prevent separation and improve the texture and mouthfeel. It also is used as a binder and adhesive in food products. It helps ingredients stick together and provides cohesiveness. It is often used in meat products, such as sausages and burgers, to improve their texture and prevent them from crumbling (Setia et al., 2010).

Gum karaya can be used as a fat replacer in certain food formulations. It can mimic the functionality of fat by providing mouthfeel and texture while reducing the overall fat content of the product. This is particularly useful in low-fat food products. It is a good source of dietary fibre. It can be added to food products to increase their content, which can have various health benefits. It aids digestion, promotes capacity, and helps regulate blood sugar levels. Gum karaya has freeze-thaw stability, which means it can withstand multiple cycles of freezing and thawing without losing its properties. This makes it suitable for frozen food products, such as ice creams and frozen desserts, where stability during storage and transportation is crucial.

12.3.2 Pharmaceutical Applications

Gum karaya is used in the pharmaceutical industry as a binder, thickener, and emulsifier in various formulations. It is often utilized in tablets, capsules, syrups, and suspensions to improve their consistency and stability. It is rich in dietary fibre, which can provide various health benefits. It aids in digestion, promotes bowel regularity, and may help reduce cholesterol levels. As a result, it is used as a dietary fibre supplement in some dietary supplements.

Gum karaya is often used as a binder in tablet formulations to hold the active ingredients together. It provides cohesiveness and helps maintain the integrity degradation of sensitive drugs and in enhancing their shelf life. It can be incorporated into drug delivery systems to achieve controlled release of medications. It forms a gel-like matrix that gradually releases the active ingredients, allowing for sustained drug release over an extended period. This is particularly useful for drugs that require prolonged action within the body.

Gum karaya has excellent emulsifying properties, making it useful in the formulations of emulsions, creams, and ointments. It helps stabilize oil-in-water emulsions by preventing the separation of the oil–water phases. Additionally, it acts as a suspending agent, preventing the setting of solid particles in liquid formulations. Due to its high viscosity, gum karaya is employed as a thickening agent in pharmaceutical formulations. It imparts a desirable texture and consistency to oral suspensions, syrups, and gels. By adjusting the concentrations of gum karaya, the viscosity of the formulations can be controlled to optimize its flow and ease of administration.

Gum karaya possesses mucoadhesive properties, meaning, it can adhere to mucous membranes. This characteristic is advantageous for oral drug delivery systems, as it enables prolonged contact of the drug with the absorption site, enhancing drug absorption and bioavailability. It is considered a safe and inert excipient, making it suitable for use in oral pharmaceutical formulations. It is used as a filler or diluent on tablets, capsules, and granules, providing bulk and aiding in the manufacturing process.

12.3.3 Textile Industry

Gum karaya has been historically used in the textile industry as a sizing agent. It provides stiffness and body to fabrics, making them easier to handle during weaving and reducing yarn breakage. It helps create a consistent and smooth paste that can hold the dyes and pigments. By improving the viscosity of the printing paste, gum karaya enables better control over the printing process and enhances the sharpness of the printed patterns. Sizing is the process of applying a protective coating to the fabric to improve its handling and weaving characteristics. Gum karaya is employed as a sizing agent due to its excellent adhesive properties. It helps increase the strength, flexibility, and smoothness of the fabric during weaving. Gum is utilized in textile finishing processes to enhance the properties of the fabric (Dai, 1994). It can be applied as a natural binder or as an additive in finishing formulations to improve fabric softness, wrinkle resistance, and crease recovery. The gum forms a film on the fabric surface, providing a protective layer and improving its overall appearance. It is often used as a thickener and binder in textile dyeing processes. It helps control the migration of dyes by preventing bleeding or spreading of colour in the fabric. By enhancing the adhesion of dyes to the textile fibres, it promotes colour fastness and improves the brightness of dyed fabric (Mittal et al., 2015). Gum possesses emulsifying properties, making it useful in textile emulsions. These are mixtures of two or more immiscible liquids, such as oil and water, stabilized with an emulsifier. It can stabilize these emulsions, ensuring the uniform dispersion of substances like oils, waxes, or pigments in water-based textile formulations.

12.3.4 Wastewater Treatment

The wastewater generated from gum karaya processing may contain organic matter, suspended solids, and potentially harmful chemicals used during extraction or purification processes. It is essential to treat this wastewater before discharging it into the environment to minimize its impact on ecosystems and human health. These are some possible roles and methods for treating wastewater. Wastewater treatment processes typically begin with the removal of suspended solids through physical methods, such as sedimentation and filtration. This step helps separate solid particles from the liquid fraction (Raj et al., 2021). Gum karaya wastewater can undergo biological treatment, where microorganisms degrade organic matter present in the wastewater. Biological processes, such as activated sludge or biofilm reactors, can be employed to break down and convert organic pollutants into less-harmful substances (Bashir et al., 2020). Depending on the specific contaminants in the wastewater, chemical treatment methods may be required. Coagulation and flocculation processes can be used to aggregate fine particles and facilitate their removal. Chemical additives like coagulants and flocculants aid in the separation of solid and liquid fractions (Mittal et al., 2015; Panic and Velickovic, 2014). Gum karaya wastewater might have acidic or alkaline pH levels. Adjusting the pH to neutral levels is often necessary before proceeding with subsequent treatment steps, as extreme pH values can hinder the efficiency of biological or chemical treatment processes. It has an acidic nature value of 4.5. If the wastewater contains persistent organic pollutants or high levels of dissolved solids, advanced treatment techniques may be needed. These can include processes such as activated carbon adsorption, membrane filtration, or advanced oxidation processes to further remove contaminants and ensure the water meets the required quality standards. Once the wastewater has undergone appropriate treatment, the effluent can be discharged into surface water bodies, subject to compliance with local environmental regulations. Alternatively, if treated to a suitable quality, it can be reused for non-potable purposes, such as irrigation or industrial processes, helping conserve freshwater resources.

12.3.5 Pharmacological Uses

It has been traditionally used in Ayurvedic medicine for its potential medicinal properties. It is believed to have anti-inflammatory, antidiarrheal, and antiulcer properties. However, further research established its efficacy and safety for therapeutic purposes. It has been traditionally used for various medicinal purposes due to its unique pharmacological properties. It is used for laxative properties and is often used to relieve constipation. It increases the bulk of the stool, promoting regular bowel movements and easing the passage of faeces. It is rich in soluble dietary fibre, which can help improve digestion and regulate bowel movements. It can be used as dietary supplement to increase fibre intake and promote gastrointestinal health. Some studies suggest that gum karaya possesses anti-inflammatory properties, including its potential anti-inflammatory effects. However, it is important to note that scientific research on gum karaya anti-inflammatory properties is limited, and more studies are needed to fully understand its potential benefits. These studies suggest that certain gums, including gum karaya, possess anti-inflammatory properties. However, it is important to keep in mind that the specific effects of gum karaya may differ from those of other gums due to variations in composition and bioactive compounds. It has been used in traditional medicine to soothe inflamed tissues, such as in the case of sore throat or oral ulcers. It is also traditionally applied topically to promote wound healing. It forms a protective layer over the wound, helping to maintain a moist environment that aids in healing and prevents infection. Polysaccharide fraction exhibited significant anti-inflammatory activity by reducing the level of inflammatory markers, such as tumour necrosis factor-alpha (TNF-α) and interleukin-6 (IL-6) in the animals (Raj et al., 2021). Gum karaya shows anti-inflammatory activity in rats with induced paw oedema. It shows that gum karaya exhibited significant anti-inflammatory effects by reducing the paw oedema and inhibiting the activity of inflammatory enzymes, such as myeloperoxidase. As with any natural product or supplement, it is always advisable to consult with a healthcare professional before using gum karaya for its potential anti-inflammatory effects. They can provide personalized advice and help assess whether it is appropriate for your specific health condition.

12.3.6 Other Applications

Gum karaya has emulsifying properties, which means it can help stabilize emulsions mixture of oil and water. It is commonly used as a food additive in various processed foods and beverages to prevent separation of ingredients and improve texture. In pharmaceutical applications, gum karaya is utilized in the pharmaceutical industry as an excipient in tablet formulations. It can be used as a binder, disintegrant, and filler in tablet manufacturing, providing structural integrity and aiding in drug release. It is used in dental care products, including toothpaste and mouthwash, due to its adhesive and emulsifying properties. It can help enhance the stability and consistency of dental formulations.

12.4 LIVELIHOODS: CASE STUDY

12.4.1 A Journey towards Sustainable Tribal Livelihoods – Kovel Foundation (by Dr. Krishna Rao)

The forest areas of Eastern Ghats in Andhra Pradesh, India, is mainly inhabited by tribal communities. The tribals are dependent on agriculture, non-timber forest produce (NTFP) collection, wage labour, etc. for their livelihoods. They collect various NTFP products like gums, resins, tamarind, amla, honey, soap nuts, mahua flowers, seeds, tubers, etc. to meeting their consumption and sale. Gums and resins are perhaps the most widely used and traded NTFP products other than those items which are consumed directly as food, fodder, and medicine. Use of gums and resins for domestic consumption and sale to earn some cash is very common among forest-dwelling communities, particularly the tribals. Among the gums, gum karaya is one of the key resources for generating a perennial source of income for thousands of families that are residing in the forests. Gum pickers follow destructive methods of collection and processing of gum karaya as they are not fully aware of scientific practices and because of a lack of required toolkits, leading to destruction of gum trees, thereby lowering returns to the dependent families. Keeping these issues in view, the Kovel Foundation, in collaboration with Tribal Cooperative Marketing Federation of India Ltd. (TRIFED) and Girijan Cooperative Corporation Ltd. (GCC), has designed a training programme for the tribal gum pickers on scientific tapping of gum karaya in Andhra Pradesh State, India, with an objective of protecting gum trees by improving the quantity and quality of gum collected.

The Kovel Foundation has trained 7,500 families in the scientific collection of gum karaya, and in this process, tribals have been trained. The Kovel Foundation has brought out a short film, 'Gum Karaya – A sustainable Livelihood' (*www.youtube.com/watch?v=SPBbODcCBnI*), giving details of this training. The Kovel Foundation, a trust of indigenous people for research on sustainable technologies, is a state-level apex body with 253 collectives of NTFP representing 8,000 gum-collecting families across the state of Andhra Pradesh. *Kovel* is the name of gum karaya in the Telugu language. Kovel Foundation's main goal is to promote sustainable livelihood, especially forest produce, and organize community collective, capacity building, value addition, and market linkages besides imparting training tribals in scientific extraction of gum karaya.

Kovel Foundation has worked since 1990 to improve the quality and quantity of gum karaya through regular intensive and refresher training programs for pickers. The training programs are organized for 25 to 30 gum pickers at any one time. They are trained on appropriate methods of gum collection, grading, and storage. Issues are discussed during these sessions, along with appropriate solutions, and also supplying toolkits to gum pickers free of charge in order to encourage the adoption of scientific practices to maintain quality standards. They were given awareness, and quality control was undertaken at all stages, from tapping, collection, and handling by the tribals.

Due the capacity building, quality was enhanced, and so higher prices were received from exporters. The grade I gum was obtained from tribals at Rs.30 per kg in 1989 to Rs. 40 in May 1992, Rs. 60 in April 1993, Rs. 80 in November 1993, Rs. 90 in May 1994, and Rs. 100 in April 1996. This enhanced quality benefit was shared by 8,500 tribal families. The benefit was to corporations and exporters too. Most importantly, the life span of gum karaya is enhanced by 20 to 25 years – a true win–win solution to a difficult problem (Gautami et al., 1992).

12.5 CONCLUSION

It is important to note that while gum karaya has a long history of traditional use, scientific research on its pharmacological effects is limited. Gum karaya will be an important source of livelihood with proper research, capacity building, and training. NTFPs and on-time intervention through scientific and technical knowledge, value addition, and enhanced competitiveness of such products lead to empowerment of the tribals. It has also resulted in the sustainable use of forest resources, helping in alleviating poverty amongst the poor tribals.

REFERENCES

Anderson, D.M.W. 1989. Evidence for the safety of Gum Karaya (*Sterculia* spp.) as a food additive. *Food Addit. Contam.* 6(2):189–199.

Bashir, I., F. Lone, R.A. Bhat, S.A. Mir, Z.A. Dar and S.A. Dar. 2020. Concerns and threats of contamination on aquatic ecosystems. In *Bioremediation and biotechnology*. (pp. 1–26). Springer.

Dai, M. 1994. The effect of zeta potential of activated carbon on the adsorption of dyes from aqueous solution: I. The adsorption of cationic dyes: Methyl green and methyl violet. *J. Colloid Interface Sci.* 164:223–228.

Dhiman, M., A. Singh and M.M. Sharma. 2019. A review on *Sterculia urens* Roxb.: A boon to the livelihood for tribal people and industry. *Industr. Crops Prod.* 130:341–351.

Gautami, S. and R.V. Bhat, 1992. *A monograph on Gum Karaya*. National Institute of Nutrition (pp. 157–165).

Jagtap, S.D., S.S. Deokule and S.V. Bhosle. 2006. Some unique ethnomedicinal uses of plants used by the Korku tribe of Amravati district of Maharashtra, India. *J. Ethopharmacol.* 107:463–469. https://doi.org/10.1016/j.jep.2006.04.002

Jain, A., S.S. Katewa, P.K. Galav and P. Sharma. 2005. Medicinal plant diversity of Sitamata wildlife sanctuary, Rajasthan, India. *J Ethnopharmacol.* 102:143–157.

Kala, C.P. 2011. Indigenous uses and sustainable harvesting of trees by local people in Pachmarhi Biosphere Reserve of India. *Int. J. Med. Arom. Plants.* 2011: 153–161.

Kala, C.P. 2016. Important gum yielding species *Anogeissus latifolia* (Roxb.) Bedd., *Boswellia serrata* Roxb. and *Sterculia urens* Roxb.: Ethnobotany, population density and management. *Appl. Ecol. Env. Sci.* 4:61–65. https://doi.org/10.12691/aees-4-3-2

Luján-Medina, G.A., J. Ventura, A.C.L. Ceniceros, J.A. Ascacio, D.B.V. Valdes, and C.N. Aguilar. 2013. Karaya gum: general topics and applications. *Macromolecules Indian J.* 9:111–116.

Mittal, H., A. Maity and S.S. Ray. 2015. Synthesis of co-polymer-grafted gum karaya and silica hybrid organic-inorganic hydrogel nanocomposite for the highly effective removal of methylene blue. *Chemical Eng. J.* 279:166–179.

Oak, G., P. Kurve, S. Kurve and M. Pejaver. 2015. Ethno-botanical studies of edible plants used by tribal women of Thane District. *J Med. Plants Stud.* 3:90–94.

Omkar, K., S. Suthari. S. Alluri, A. Ragan and V.S. Raju. 2012. Diversity of NTFPs and their utilization in Adilabad district of Andhra Pradesh, India. *J. Plant Studies.* 1(1):33.

Panduranga, R.M., S. Prasanthi and S.T. Reddi. 2011. Medicinal plants in folk medicine for women's diseases in use by Konda reddis. *Indian J. Trad. Knowl.* 10:563–567.

Panic, V.V. and S.J. Velickovic. 2014. Removal of model cationic dye by adsorption onto poly (methacrylic acid)/zeolite hydrogel composites: Kinetics, equilibrium study and image analysis. *Separation Purif. Technol.* 122:384–394.

Raj, V., J.H. Lee, J.J. Shim and J. Lee. 2021. Recent findings and future directions of grafted gum karaya polysaccharides and their various applications: A review. *Carbohydr Polym.* https://doi.org/10.1016/j.carbpol.2021.117687

Rao, D.S., G.M.M. Rao and P.P. Murthy. 2016. Diversity and indigenous uses of some ethnomedicinal plants in Papikondalu wildlife sanctuary, Eastern Ghats of Andhra Pradesh, India. *Am. J. Ethnomed.* 3:6–25.

Ratnam, K.V. and R.R.V. Raju. 2005. Folk medicine used for common women ailments Adivasis in the Eastern Ghats of Andhra Pradesh. *Indian J. Trad. Knowl.* 4:267–270.

Sahu, P., P.S. Pisalkar, S. Patel and P. Katiyar. 2019. Physico-chemical and rheological properties of Karaya Gum (*Sterculia urens* Roxb.). *Int. J. Curr. Microbiol. App. Sci.* 8(4): 672–681. https://doi.org/10.20546/ijcmas.2019.804.072

Setia, A., S. Goyal and N. Goyal. 2010. Applications of gum karaya in drug delivery systems: A review on recent research. *Der Pharmacia Lettre.* 2(5):39–48.

Shekarforoush, E., H. Mirhosseini, B.T. Amid, H. Ghazali, K. Muhammad, M.Z.I. Sarker, and M. Paykary. 2016. Rheological properties and emulsifying activity of gum Karaya (*Sterculia urens*) in aqueous system and oil in water emulsion: heat treatment and microwave modification. *Intern. J. Food Propert.* 19(3):662–679.

Vaidyanathan, D., M.S. Senthilkumar and M.G. Basha. 2013. Studies on ethnomedicinal plants used by Malayali tribals in Kolli hills of Eastern Ghats, Tamil Nadu, India. *Asian J. Plant Sci. Res.* 3:29–45.

Index